Quantitative Structure–Activity Relationship

Quantitative Structure–Activity Relationship

A Practical Approach

Siavoush Dastmalchi
Maryam Hamzeh-Mivehroud
Babak Sokouti

CRC Press
Taylor & Francis Group
Boca Raton London New York

CRC Press is an imprint of the
Taylor & Francis Group, an **informa** business

CRC Press
Taylor & Francis Group
6000 Broken Sound Parkway NW, Suite 300
Boca Raton, FL 33487-2742

Printed on acid-free paper

International Standard Book Number-13: 978-0-8153-6209-8 (Hardback)

Library of Congress Cataloging-in-Publication Data

Names: Dastmalchi, Siavoush, 1966- author. | Hamzeh-Mivehroud, Maryam, 1981- author. | Sokouti, Babak, 1978- author.
Title: Quantitative structure - activity relationship : a practical approach / Siavoush Dastmalchi, Maryam Hamzeh-Mivehroud and Babak Sokouti.
Description: Boca Raton : CRC Press, [2018] | Includes bibliographical references and index.
Identifiers: LCCN 2017059795| ISBN 9780815362098 (hardback : alk. paper) | ISBN 9781351113076 (ebook)
Subjects: LCSH: Drugs--Structure-activity relationships. | QSAR (Biochemistry) | Drugs--Design.
Classification: LCC RM301.42 .D36 2018 | DDC 615.1/9--dc23
LC record available at https://lccn.loc.gov/2017059795

Visit the Taylor & Francis Web site at
http://www.taylorandfrancis.com

and the CRC Press Web site at
http://www.crcpress.com

Contents

Foreword

THE IMPORTANCE OF QUANTITATIVE STRUCTURE–ACTIVITY AND STRUCTURE–PROPERTY RELATIONSHIPS

There has been always an interest in the relationship between chemical structure and chemical and biological properties for a very long time. Although the modern science of quantitative structure–activity relationship (QSAR) is considered by many to have started with a seminal publication by Corwin Hansch and coworkers (1962), many people had been laying the groundwork from the early part of the nineteenth century (Dearden 2016). Even more than two millennia ago, the Roman poet and philosopher Lucretius wrote in his *De Rerum Natura* (*On the Nature of Things*) (Leonard 2016):

> *We see how quickly through a colander*
> *The wines will flow; how, on the other hand,*
> *The sluggish olive-oil delays: no doubt,*
> *Because 'tis wrought of elements more large,*
> *Or else more crook'd and intertangled.*

Lucretius was implying that viscosity was a function of molecular size, shape, and length, which, for a time around 60 BC, was a remarkably perspicacious insight.

Following the work of Hansch et al. (1962), QSAR studies increased in number, and publications reached over 1400 a year by 2014 (Dearden 2016), as found using the search terms *QSAR*, *QSPR*, and *quantitative structure–activity relationship*. However, some of those publications contained errors. As pointed out by Gramatica et al. (2012), QSAR modeling is not just "push a button and find a correlation." Dearden et al. (2009) reported a total of 21 different types of error made in QSAR studies. Some of those errors are still being made, partly at least because to be a good QSAR practitioner, one should ideally be a physical chemist, an organic chemist, a biochemist, a biologist, a pharmacologist, a toxicologist, a mathematician, and a statistician!

It is clear from the above that despite modern QSARs being over 55 years old, there is still a need for definitive "how to do it" guidance on the subject. This book provides that guidance, with a clear and logical layout, starting with an introductory chapter on what QSAR is; and moving on to databases and data sets, molecular descriptors, and descriptor selection; how to build and validate QSAR models; and finishing with a comprehensively worked example.

This book should be useful and valuable for all QSAR practitioners, whether new to the science or with wide experience. Students will also find it helpful.

REFERENCES

Dearden, J.C., M.T.D. Cronin, and K.L.E. Kaiser. 2009. How not to develop a quantitative structure-activity or structure-property relationship (QSAR/QSPR). *SAR and QSAR in Environmental Research* 24 (7): 263–274.

Dearden, J.C. 2016. The history and development of quantitative structure-activity relationships (QSARs). *International Journal of Structure-Property Relationships* 1 (1): 1–43.

Gramatica, P., S. Cassani, P.P. Roy, S. Kovarich, C.W. Yap, and E. Papa. 2012. QSAR modelling is not "push a button and find a correlation": A case study of toxicity of (benzo-) triazoles on algae. *Molecular Informatics* 31 (11–12): 817–835.

Hansch, C., P.P. Maloney, T. Fujita, and R.M. Muir. 1962. Correlation of biological activity of phenoxyacetic acids with Hammett substituent constants and partition coefficients. *Nature* 194 (4824): 178–180.

Leonard, W.E. 1916. *Lucretius: De Rerum Natura*. New York: E.P. Dutton.

John C. Dearden

Preface

Quantitative structure–activity relationship (QSAR) analyses, introduced more like to what we know of it today by the pioneering work of Corwin H. Hansch, aim to correlate the structural features of molecules to their biological activities using mathematical models. Since the inception, QSAR studies have impressive effects on drug design and discovery processes. Such analyses are commonly used by researchers who are mainly involved in medicinal chemistry-based projects. The integration of QSAR studies in medicinal chemistry is so strong that almost all well-known textbooks and monographs in medicinal chemistry have devoted specific chapter(s) to this topic. The same is also true, not surprisingly, for the teaching of medicinal chemistry. The idea of writing a book on QSAR was intrinsic to us due to the nature of being in the academia and teaching/working with students in the field. The current book was organized based on the natural flow of conducting a QSAR study, starting from the data set preparation, all the way through molecular descriptor selection, model building, and evaluation. At the end of the book, a practical example was provided for further discussion on the different steps of a QSAR analysis. The focus of the current volume is on two-dimensional QSAR, but most of the steps explained herein are equally applicable to three-dimensional studies as well. It took us about six months from the idea of writing the book to have the final draft ready in hand for production and along the way many people have helped us directly or indirectly. We would like to express our greatest gratitude to the people who have helped and supported us throughout this project, but the support we received from Prof. John C. Dearden was incomparable. We also gratefully acknowledge the help of Ms. Zoha Khoshravan Azar for the preparation of the data set used in Chapter 8.

Siavoush Dastmalchi
Biotechnology Research Center & School of Pharmacy
Tabriz University of Medical Sciences
Tabriz, Iran

Maryam Hamzeh-Mivehroud
Biotechnology Research Center & School of Pharmacy
Tabriz University of Medical Sciences
Tabriz, Iran

Babak Sokouti
Biotechnology Research Center & School of Pharmacy
Tabriz University of Medical Sciences
Tabriz, Iran

The MathWorks, Inc.
3 Apple Hill Drive
Natick, MA 01760-2098 USA
Tel: 508 647 7000
Fax: 508-647-7001
E-mail: info@mathworks.com
Web: www.mathworks.com

Authors

Siavoush Dastmalchi graduated as a doctor of pharmacy from Tabriz University of Medical Sciences (TUMS), Tabriz, Iran. Then he moved to Sydney where he received his PhD from the Faculty of Pharmacy at the University of Sydney, Sydney, Australia, in 2002. Since then he has been working as a full academic in the Medicinal Chemistry Department at the School of Pharmacy, TUMS, teaching medicinal chemistry, instrumental drug analysis, and bioinformatics to graduate and postgraduate students. He is currently the director of the Biotechnology Research Center and in the meantime heads the Department of Medicinal Chemistry at TUMS. He leads a research team mainly involved in molecular modeling, structural biology, and chemo-bioinformatics for their application in drug discovery.

Maryam Hamzeh-Mivehroud is an associate professor of Medicinal Chemistry who graduated as doctor of pharmacy from Tabriz University of Medical Sciences, Tabriz, Iran, in 2004 and received her PhD from the same university in 2011. Since then she has been working as a full academic member in the Medicinal Chemistry Department at the School of Pharmacy and teaches medicinal chemistry at the undergraduate and postgraduate levels. Her main research interests are focused on quantitative structure–activity relationship (QSAR) and molecular modeling in the field of drug design and discovery.

Babak Sokouti is an assistant professor of bioinformatics in the Biotechnology Research Center at Tabriz University of Medical Sciences Tabriz, Iran, with over 19 years of experience in IT technical management and consulting, including managing and maintaining sophisticated network infrastructures. He has obtained bachelor of science in electrical engineering with a specialization in Control from Isfahan University of Technology, Isfahan, Iran; a master of science in Electrical Engineering with a specialization in electronics (with background of biomedical engineering) from Tabriz Branch, Islamic Azad University, Tabriz, Iran; a master of science in information security with Distinction from Royal Holloway University of London, London, UK; and obtained PhD in bioinformatics from Biotechnology Research Center, Tabriz University of Medical Sciences, Tabriz, Iran. His research interests include cryptographic algorithms, information security, network security and protocols, image processing, protein structure prediction, and hybrid intelligent neural network systems based on genetic algorithms.

1 QSAR at a Glance

Design is not just what it looks like and feels like. Design is how it works.

Steve Jobs

The increasing growth of data in the era of information technology requires accelerated data availability for conducting computational and statistical processes to achieve useful information through data mining approaches. There is no "single" methodology for gathering the information due to the diverse nature of different problems. Hence, different specific and customized algorithms have been developed in various fields of science to achieve the desired goal. Quantitative structure–activity relationship (QSAR) is a chemoinformatic technique, which involves data mining and has proved to be helpful in accelerating the process of drug design and discovery. QSAR is defined as a mathematical equation that correlates the biological activities of the compounds to their structural features. The generic term for QSAR analysis can be expressed as

Predicted biological activity = Function (structural features)

The pioneering QSAR analysis most akin to what we know today dates back to the early 1960s, in which a relationship was established between partition coefficients of compounds and their biological endpoints by Hansch et al. (1962). Since then numerous technical and theoretical advances have been introduced into the field with no apparent end in sight. The main objectives of QSAR studies are (1) prediction of biological activity of the compounds, (2) lead optimization, (3) design of new active compounds, (4) prediction of risk assessments and toxicity, (5) modeling of pharmacokinetic and pharmacodynamic profile of new chemical entities, and (6) discovery of compound(s) with desired biological activity by screening chemical databases or virtual libraries (Gramatica 2007, Golbraikh et al. 2012, Kirchmair et al. 2015, Roy et al. 2015, 2015b, Wang et al. 2015).

Every QSAR analysis needs to follow a workflow consisting of several computational and statistical steps as shown in Figure 1.1. Briefly, these steps comprise data collection and preparation, calculation and preprocessing of molecular parameters, data sets generation (train and test sets), descriptor selection, model building, internal and external validation, and model development. Moreover, the mentioned steps can be followed by an experimental validation.

A representative example of a QSAR model for the relative potencies (log *RP*) of antihyperglycemic effects of sulfonylurea drugs is shown in the following equation:

$$\log RP = -(1.126 \pm 0.289) + (0.161 \pm 0.019)a\_hyd - (0.214 \pm 0.092)HS8 \qquad (1.1)$$

$$n = 13, R^2 = 0.865, Q^2 = 0.744, s = 0.290, F = 35.1, \text{ all } p \text{ values} \leq 0.04$$

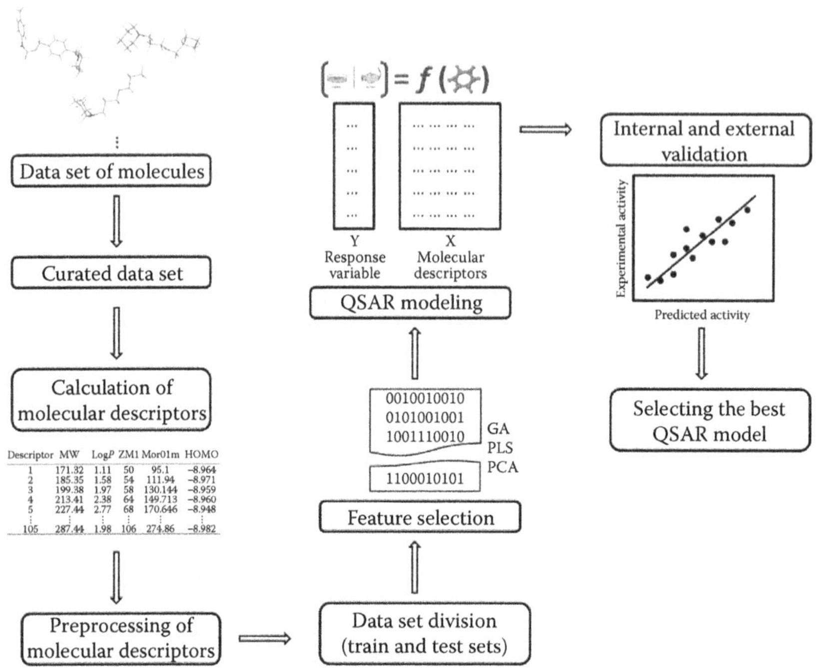

FIGURE 1.1 A schematic representation of QSAR workflow.

where:

a_hyd is the number of hydrophobic atoms

*HS*8 is the atom-level E-state value for hydrogen atom number 8 related to its hydrogen bonding capability

n is the number of compounds

R^2 is the squared correlation coefficient of determination

Q^2 demonstrates leave-one-out cross-validated coefficient of determination

s and *F* parameters are standard error of prediction and Fisher statistic, respectively (Dearden et al. 2015)

There are several important rules for developing a QSAR model capable of predicting activity/property/toxicity of a new compound falling in the applicability domain (AD) of the model. In 2002 in a meeting held in Setubal, Portugal, some regulatory principles for validating QSAR models were proposed by QSAR experts from all over the world, which were later adopted in the 37th Joint Meeting of the Chemicals Committee and Working Party on Chemicals, Pesticides and Biotechnology held in November 2004 by members of Organization for Economic Co-operation and Development (OECD). They agreed on five principles proposed as a guideline for regulatory purposes to be considered in any QSAR model development (OECD 2004). These principles are as follows:

1. A defined endpoint
2. An unambiguous algorithm
3. A defined domain of applicability
4. Appropriate measures of goodness-of-fit, robustness, and predictivity
5. A mechanistic interpretation, if possible

A comprehensive description for each item of OECD principles is available in the Appendix A.

In the current book, different steps of a QSAR analysis workflow are discussed in detail along with potential limitations and challenges based on OECD rules in the next chapters. In addition, a practical example is provided in Chapter 7 as a "hands on practice" example of generating a QSAR model for a set of 105 histamine H3 receptor antagonists.

2 Database and Dataset

Data is the new science. Big Data holds the answers.

Pat Gelsinger

2.1 INTRODUCTION

"What is the starting point for a QSAR-based project?", "Where do the relevant data come from?", "What are the features of reliable data?" are the types of questions one may think of at the beginning of a QSAR study. To answer these questions, understanding the fundamental concepts of QSAR-based studies is essential. In this chapter, it has been tried to cover most of the considerations dealing with data, data collection, databases, and other possible related problems such as errors and biases in data.

Data as an error-prone object in any computational environments should be handled with good care to construct a possible error-free data set for further analysis.

Bioactive molecules exert biological activity on the desired targets. The data related to these compounds originate from the experimental studies. Biological activities are quantitative measurements with broad definition in which the interaction(s) between molecule of interest and its corresponding receptor plays a central role. Broadly speaking, the experimental nature of these measurements varies from test to test and is divided into four bioassay classes including biochemical, cell-based, animal, and human assays. Biochemical assays, also called *in vitro* assays, are the first, the simplest, and the cheapest methods considered for initial assessment of compounds' efficacy toward the desired target of interest. Examples of these bioassays are affinity (e.g., equilibrium dissociation constant K_d) and inhibitory (e.g., K_i, IC_{50}) data. Other *in vitro* bioassays, namely cell-based assays, are performed using living cultured cells that are more complex, requiring more expensive procedures than the former assays. An enzyme inhibition inside the cells can be regarded as an example. *In vivo* tests of compounds refer to biological evaluation carried out within an animal or a human being.

For achieving a good QSAR model, there are some recommendations on the total number of compounds and the range of biological activity and standard deviation values (Gedeck et al. 2006, Hamzeh-Mivehroud et al. 2015, Scior et al. 2009). Practically, the number of studied compounds should not be too less for some obvious reasons, such as chance correlation and overfitting or too much to avoid computational restrictions (Tropsha 2010), which may not be a real concern for most of the researchers these days. Based on a suggestion by Scior et al., at least 20 molecules are required for a QSAR analysis (Scior et al. 2009). The minimum acceptable range of biological activities in a data set is one order of magnitude, but usually to have a good intrapolation of activities, a range about 3.0 folds in logarithmic scale is recommended (Gedeck et al. 2006, Hamzeh-Mivehroud et al. 2015, Scior et al. 2009). Moreover, the standard deviation also needs to satisfy the criterion of at least 1.0 logarithmic unit (Gedeck et al. 2006). Two out of five rules provided by the Organization for Economic Co-operation and Development (OECD) refer to data characteristics, which are the initial requirements to be satisfied by a data set of compounds (OECD 2004). These two rules, namely "a defined endpoint" and "a defined domain of applicability", are the first and the third bullets of the guideline, respectively. These two rules can be violated by a variety of issues including data heterogeneity, inappropriate endpoint data, undefined applicability domain (AD), unacknowledged omission of data points, inadequate data, replication of compounds, and too narrow range of endpoint values (Cherkasov et al. 2014, Dearden et al. 2009). Some biological data used for the development of a QSAR model should be derived from the same experimental procedure. For example, the collected biological activities or endpoint values should be obtained from the same species using the same protocols (Cherkasov et al. 2014, Dearden et al. 2009). The units utilized for defining biological activities should obey the International System of Units (SI), known as metric system. For instance, all the values for IC_{50} should be expressed in molar unit not weight unit.

AD is another criterion that should be considered in the development of QSAR models. This means that, if a structurally similar compound falls into the chemical space of the trained model, the predicted biological activity should also be in the range of biological activities used in the QSAR model for a highly reliable prediction (Cherkasov et al. 2014, Dearden et al. 2009, Roy et al. 2015). In other words, theoretical determination of predictive power of a QSAR model should be taken into consideration by means of AD tools using the descriptors and biological activities in train and test sets (Roy et al. 2015). The data used for building a QSAR model are generally extracted from experimentally published data in the literature after properly curating the data—for example, by removing outliers to enhance the QSAR-model-predictive capability. This omission of data points should be supported by a good explanation.

Besides the above-mentioned issues dealing with data curation, some other issues can also be taken into consideration such as incorrect Chemical Abstracts Service (CAS) numbers, chemical names, and structures (Cherkasov et al. 2014, Dearden et al. 2009). The replicated chemical structures, which could be present frequently in both train and test data sets, may falsely increase the predictive performance of QSAR models. The source of these replications may include the same structures with different CAS numbers or names as well as different biological activities (Cherkasov et al. 2014, Dearden et al. 2009). How to select a suitable data sets for developing a QSAR model in terms of biological activity range is still debatable. Use of too narrow range of endpoint data makes the QSAR model unreliable. The challenging issues dealing with QSAR modeling should be carefully addressed from the very first stage of data collection, as a very minor error in manipulation of the data set at this stage will be distributed across the entire QSAR modeling procedure and will lead to a poor performance of the obtained model.

2.2 DATA SOURCES

2.2.1 Database Servers

Over the past decades, astronomical generation of various types of information whether they are embedded as the web-based or stand-alone databases provides useful means for many studies especially in the realm of computational chemistry. The QSAR analysis is one of those predictive methods, which highly depends on "Data". The input data, the central part of any chemoinformatics study, can be extracted from either available literature or databases. These types of data typically consist of molecular structures along with their corresponding biological activities (Gozalbes and Pineda-Lucena 2011, Zhou 2011). Publicly or commercially available databases provide means for maintenance, storage, manipulation, and extraction of large amount of compounds and related data (Gozalbes and Pineda-Lucena 2011, Zhou 2011). There are several databases reported in the literature; some of which are listed in Table 2.1 as the representatives, which are commonly used in QSAR studies. The information on these databases in the table includes their availability, accessibility of chemical structures and biological activities, and the website addresses. From chemical structural point of view, these databases contain highly diverse molecules. Moreover, the databases are searchable either by entering the chemical name or manually drawing the structure of the compounds. The current list of database servers in the table is divided into thirteen publicly available and nine commercial databases. The Available Chemicals Directory (ACD) database is the gold standard source for retrieving chemical information. This collection as a structure-searchable database consists of more than 10 million unique chemicals as well as 3D models without their corresponding biological activities. The SPRESIweb v2.14, CAS REGISTRY, and iResearch Library Collection also contain more than five million unique compounds from which iResearch Library Collection reaches almost sixty million—all of which are from commercial sector. DrugBank v4.5 (available from OMX Inc. with 8206 compounds) (Knox et al. 2011, Law et al. 2014, Wishart et al. 2006, 2008), Comprehensive Medicinal Chemistry database (CMC) and GOSTAR GVK BIO (both containing biological activity), Screening Compounds Directory (SCD) (more than 9.7 million unique "drug-like" chemicals), World Drug Index (WDI) (with almost 80,000 drugs), Chapman & Hall/CRC Dictionary of Drugs (more than 40,000 pharmacologically active compounds) are other databases available through the web services commercially offered to customers. Table 2.1 also tabulates several freely accessible databases that are extensively used in QSAR projects. Some of the free databases do not provide biological activity data for the entries. These are eMolecules, ChemSpider, ZINC, Physical Properties Database (PHYSPROP), ChemBioBank, SuperDrug, and MayBridge consisting of 6, 52, and 35 million; 41,000; more than 5000; nearly 2400; and 500,000 molecules, respectively. There are also free databases that not only contain molecular structures but also their biological activities such as PubChem (89 million compounds), ChemBioNet (140,000 compounds), Open Chemical Repository Collection (200,000 compounds), ChEMBL/ChEMBL12 (nearly 2 million molecules), and BioPrint (2,500 compounds).

TABLE 2.1
Representative Overview on Databases and Their Properties

			Information		
No.	**Database**	**Availability**	**Structure**	**Activity**	**Website Address**
1	Available Chemicals Directory (ACD)	Commercial	☑	☐	http://accelrys.com/products/databases/sourcing/available-chemicals-directory.html
2	SPRESI	Commercial	☑	☐	http://www.spresi.com/
3	CAS REGISTRY	Commercial	☑	☐	http://cas.org/expertise/cascontent/registry/index.html
4	iResearch Library Collection	Commercial	☑	☐	http://www.chemnavigator.com/cnc/products/iRL.asp
5	PubChem	Free	☑	☑	http://pubchem.ncbi.nlm.nih.gov/
6	eMolecules	Free	☑	☐	http://www.emolecules.com
7	ChemSpider	Free	☑	☐	http://www.chemspider.com
8	DrugBank v4.5	Free	☑	☐	http://www.drugbank.ca
9	ZINC	Free	☑	☐	http://zinc.docking.org
10	Open Chemical Repository Collection	Free	☑	☑	https://dtp.cancer.gov/organization/dscb/obtaining/default.htm
11	ChemBioNet	Free	☑	☑	http://chembionet.de
12	ChemBioBank	Free	☑	☐	http://www.chembiobank.com
13	Physical Properties Database (PHYSPROP)	Free	☑	☐	http://www.srcinc.com/what-we-do/product.aspx?id=133
14	Comprehensive Medicinal Chemistry database (CMC)	Commercial	☑	☑	http://accelrys.com/products/collaborative-science/databases/bioactivity-databases/comprehensive-medicinal-chemistry.html
15	Screening Compounds Directory (SCD)	Commercial	☑	☐	http://accelrys.com/products/collaborative-science/databases/sourcing-databases/biovia-screening-compounds-directory.html
16	World Drug Index (WDI)	Commercial	☑	☐	http://www.daylight.com/products/wdi.html
17	Chapman & Hall/CRC Dictionary of Drugs	Commercial	☑	☐	http://dod.chemnetbase.com/tour/
18	SuperDrug	Free	☑	☐	http://bioinformatics.charite.de/superdrug/
19	ChEMBL/ChEMBL12	Free	☑	☑	https://www.ebi.ac.uk/chembl/
20	BioPrint	Free	☑	☑	http://www.cerep.fr/cerep/users/pages/ProductsServices/bioprintservices.asp
21	GOSTAR GVK BIO	Commercial	☑	☑	https://gostardb.com/gostar/
22	Maybridge	Free	☑	☐	http://www.maybridge.com/

2.2.2 Data Handling

The way of handling data extracted from any source is of great importance in data-mining projects such as QSAR-based studies. Introduction of errors can occur at different stages of data handling either during importing the data from database servers or collecting them from the literature. Therefore, final data should be carefully inspected before proceeding to the next step. Figure 2.1a illustrates the main steps of database generation in the form of a very simplified flowchart. The information stored in these databases is the major source of data available for the users that can be combined with data gathered on the basis of individual literature search and/or experimentally obtained data for preparing the data set ready to be used for the QSAR analyses (Figure 2.1b). The errors discussed earlier can happen at data set preparation steps shown in Figure 2.1b.

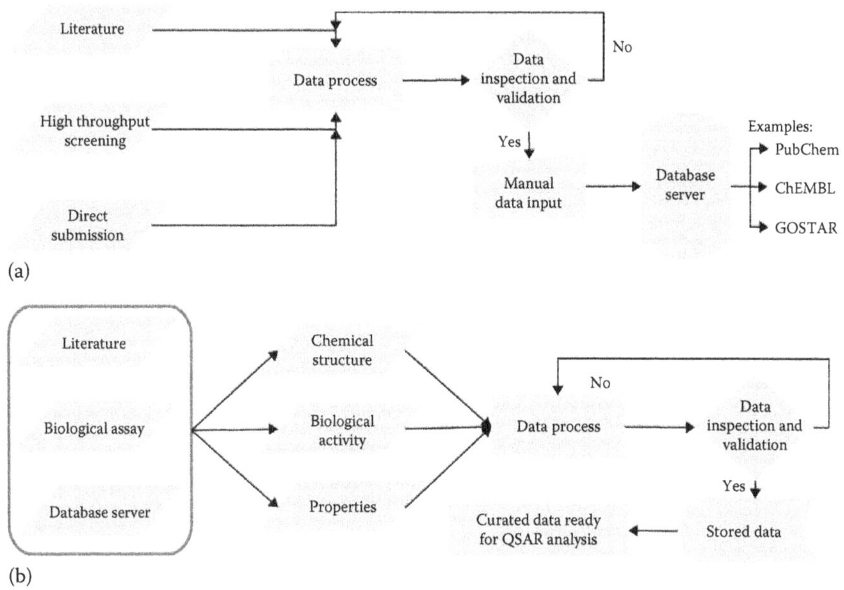

FIGURE 2.1 Process of data-handling workflow: (a) a schematic diagram shows the key steps in generation of database servers and (b) a simple flowchart shows the process of preparing data set for QSAR analysis.

3 Molecular Descriptors

3.1 INTRODUCTION

> It has been said frequently that there are three keys to the success of any QSAR model building exercise: descriptors, descriptors, and descriptors.
>
> **Tropsha (2003)**

The numerical representation of information embedded in any chemical structure is known as "molecular descriptors" (Danishuddin and Khan 2016, Gozalbes and Pineda-Lucena 2011). Considering the molecular structures of compounds, how to map their chemical structure into molecular descriptors is debatable. The molecular descriptors are experimentally and/or computationally determined and used in QSAR analyses.

For computing these parameters, whole or a fragment of a molecule is taken into account, from which whole molecule or fragment-based descriptors are generated (Danishuddin and Khan 2016, Varnek 2011). The nature of the molecular parameters varies on the basis of their types and algorithms. The parameters are often categorized as constitutional, topological, geometrical, thermodynamic, and electronic descriptors (Danishuddin and Khan 2016). Moreover, they can also be classified on the basis of the dimensionality as 0D, 1D, 2D, 3D, and others. Based on their origin, descriptors can also be divided into no-structure- (e.g., Log *P*, p*K*a, molar refractivity [MR], dipole moment, and polarizability) and structure-based descriptors that are the parameters that can be theoretically computed from the representation of molecules.

It is worth mentioning that the calculation of the values related to some of the molecular descriptors is conformation dependent. Therefore, geometry optimization should be carried out for the structures before computing such descriptors. This leads to the structures with minimum energy content. Different levels of approximation for geometry optimization are considered using quantum mechanics and/or molecular mechanics techniques (Rocha et al. 2006, Stewart 2007).

3.2 MOLECULAR DESCRIPTORS

3.2.1 Whole Molecule Descriptors

The most commonly used molecular parameters calculated for a molecule as a whole are described below using illustrative examples.

Constitutional descriptors reflect the chemical properties related to molecular composition such as molecular weight (MW), number of atoms, bond count, contribution of atom types (such as H, C, N, and O), and number of multiple bonds. Figure 3.1 illustrates a two-dimensional structure (2D) of an example compound.

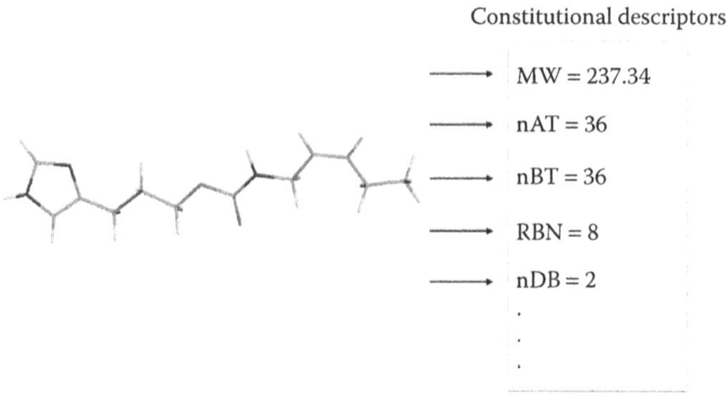

FIGURE 3.1 An example of demonstrating few constitutional descriptors calculated for compound 52 using Dragon. MW: molecular weight; nAT: number of atoms; nBT: number of bonds; RBN: number of rotatable bonds; and nDB: number of double bonds.

Another kind of descriptors derived from graph theory are called "topological descriptors" (Akutsu and Nagamochi 2013, Danishuddin and Khan 2016, Dehmer et al. 2010). These indices originate from the 2D-graph representation of compounds in which the atoms and bonds are denoted by vertices and edges, respectively. They provide information such as molecular size, overall shape, hybridization state, degree of branching, and flexibility (Danishuddin and Khan 2016, Gozalbes and Pineda-Lucena 2011, Helguera et al. 2008). The earliest example of the topological descriptors is Wiener index (Wiener 1947). This parameter counts the number of bonds between pairs of atoms (number of edges between vertex pairs). Figure 3.2 illustrates how to calculate this index based on the provided graph and distance matrix.

The other commonly used topological descriptors are Randic index (Randic 1975), Kier and Hall indices (Kier and Hall 1986), Kappa indices (Hall and Kier 2007), Zagreb index (Graovac et al. 1972), and Balaban *J* index (Balaban 1982). In Randic index, first the product of number of nonhydrogen atoms adjacent to each pair of atoms in a bond is calculated, then the sum of reciprocal of their square root is computed for defining the branching index of a molecule. A modified version of Randic index is Kier and Hall indices in which the valence connectivity and number of hydrogen atoms are introduced (Randic 1975). Kappa indices are derived from the shape of the molecule and are not related to their geometrical parameters (Hall and Kier 2007). In addition, Zagreb index is derived with the same concept of Randic index in which the total π energy of the molecular structure is considered (Graovac et al. 1972). Balaban *J* index has included the number of bonds and rings to the Randic index (Balaban 1982). For more comprehensive detail on topological indices, readers are encouraged to study the chapter written by Dearden (2017).

Geometrical descriptors are influenced by three-dimensional (3D) spatial arrangements of atoms. In other words, they are originated from the molecular conformation and atomic van der Waals areas (Danishuddin and Khan 2016, Dudek et al. 2006). Examples of these kinds of 3D descriptors are weighted holistic invariant molecular descriptors (WHIM), 3D-autocorrelation, Geometry, Topology, and Atom-Weights

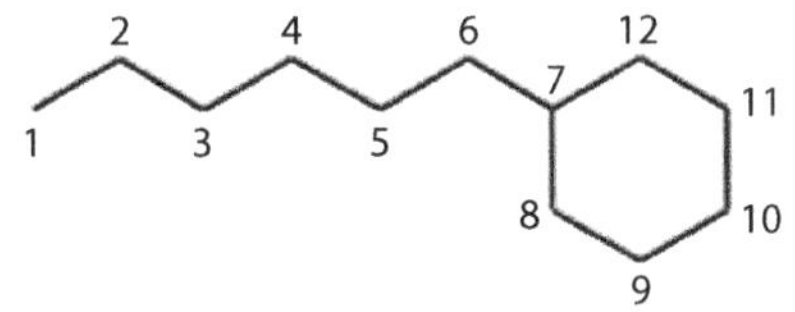

	1	2	3	4	5	6	7	8	9	10	11	12
1	0	1	2	3	4	5	6	7	8	9	8	7
2	1	0	1	2	3	4	5	6	7	8	7	6
3	2	1	0	1	2	3	4	5	6	7	6	5
4	3	2	1	0	1	2	3	4	5	6	5	4
5	4	3	2	1	0	1	2	3	4	5	4	3
6	5	4	3	2	1	0	1	2	3	4	3	2
7	6	5	4	3	2	1	0	1	2	3	2	1
8	7	6	5	4	3	2	1	0	1	2	3	2
9	8	7	6	5	4	3	2	1	0	1	2	3
10	9	8	7	6	5	4	3	2	1	0	1	2
11	8	7	6	5	4	3	2	3	2	1	0	1
12	7	6	5	4	3	2	1	2	3	2	1	0

$$W = \frac{1}{2}\sum_{i=1}^{N}\sum_{j=1}^{N} D_{ij}$$

$$W = \frac{1}{2}(60 + 50 + 42 + 36 + 32 + 30 + 30 + 36 + 42 + 48 + 42 + 36) = 242$$

FIGURE 3.2 Representation of hydrogen-suppressed molecular graph of compound 1 and distance matrix for calculating the value of the Wiener index as a simple example of topological descriptor based on the distances between atom pairs denoted by *i* and *j*.

AssemblY (GETAWAY), and 3D-molecular representation of structures based on electron diffraction (3D-MoRSE) descriptors.

WHIM descriptors contain the information about size, shape, symmetry, and atom distribution that can be categorized into two groups: (1) directional WHIM descriptors and (2) global WHIM descriptors (Todeschini and Consonni 2008a).

The other geometry-based parameters are 3D-autocorrelation descriptors (Wagener et al. 1995) calculated for a certain area containing distinct atoms and have unique values for different conformation of compounds (Danishuddin and Khan 2016).

GETAWAY descriptors, defined as the atomic influence in shape of molecular structures, are calculated using the diagonal matrix of atoms for molecular structures. In these descriptors, molecular influence matrix (MIM) is constructed from the Cartesian coordinates of atoms (i.e., *x*, *y*, and *z*) in a given special conformation for the structure (Consonni et al. 2002a,b).

Electron diffraction studies are applied for transforming the information on the basis of spatial atomic coordinates and resulted in generation of 3D-MoRSE descriptors. As revealed recently, these descriptors do not define the complexity of the molecular structures; however, they are good for describing the structure with variations in bond order and interatomic distances (Devinyak et al. 2014).

The following example is an illustrative of geometrical descriptors subjected to 3D-MorSE descriptors. The following simple electron diffraction extracted function is shown as a general formula to calculate various 3D-MorSE descriptors that can also be weighted by atomic mass, van der Waals volume, and electronegativity:

$$I(s) = \sum_{i=2}^{N} \sum_{j=1}^{i-1} A_i A_j \frac{\sin\left(s \times r_{ij}\right)}{s \times r_{ij}} \tag{3.1}$$

where:

s as an integer value shows the scattering parameter
r_{ij} is the Euclidean distance between atoms i and j
N is the total number of atoms
A_i and A_j are the features that are used for the weighting of the molecular descriptors

In this simple example, methane is considered for the calculation of Mor01u descriptor where scattering parameter is set to zero (i.e., $s = 0$ Å^{-1}) and as it is an unweighted parameter, the values for A_i and A_j are unity. So, after substitution of these values, the simplified function will be as follows (by considering that $\lim_{\theta \to 0} \frac{\sin(\theta)}{\theta} = 1$):

$$I(0) = \sum_{i=2}^{N} \sum_{j=1}^{i-1} \frac{\sin(0)}{0} = \sum_{i=2}^{N} \sum_{j=1}^{i-1} 1 = \frac{N(N-1)}{2} \tag{3.2}$$

As the total number of atoms is 5, the Mor01u value will be 10.

The second illustrative example for the calculation of Mor02u for this molecule is presented in Figure 3.3. In this instance, interatomic distances between pairs of C–H and H–H are considered and used for determining this descriptor.

No.	Atom pair	Interatomic distance, Å
1	C–H(1)	1.085
2	C–H(2)	1.085
3	C–H(3)	1.085
4	C–H(4)	1.085
5	H(1)–H(2)	1.772
6	H(1)–H(4)	1.772
7	H(2)–H(3)	1.772
8	H(3)–H(4)	1.772
9	H(2)–H(4)	1.772
10	H(1)–H(3)	1.772

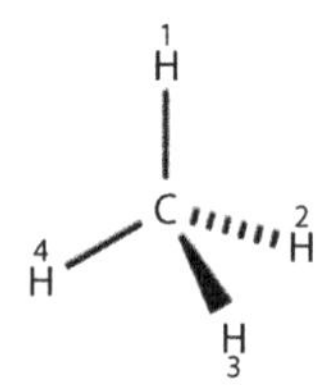

FIGURE 3.3 Parameters of methane 3D structure required for calculating $\text{MoR2u} = 4\frac{\text{Sin}(1.085)}{1.085} + 6\frac{\text{Sin}(1.772)}{1.772} = 6.578$ for 4C–H and 6H–H pairs.

The above-mentioned examples for 3D-MorSE descriptors are only the unweighted descriptors; however, these types of calculations can be performed using weighted values within a range of scattering parameters.

Thermodynamic features provide the information about the dynamics of biochemical interactions. The examples of these descriptors are lipophilicity, MR, and heat of formation that can be generated experimentally (Danishuddin and Khan 2016, Pliska 2010). The most well-known physicochemical parameter is lipophilicity (shown as Log *P*, which is the logarithm of partition coefficient value for a compound) that shows hydrophobic characteristics of the molecule. The reason for its great importance as a molecular descriptor in drug design and discovery process is related to its influence on absorption, distribution, and penetration of drugs throughout the body as well as its importance in ligand–receptor interaction. It can be measured experimentally or calculated computationally. The following equation is used for determining the lipophilicity of the studied compound as distribution between nonaqueous and aqueous solvents (Dunn et al. 1986, Kubinyi 1993).

$$P = \frac{[C]_{1\text{-octanol}}}{[C]_{\text{aqueous}}} \tag{3.3}$$

Another molecular physicochemical descriptor is MR that indicates total polarizability of a molecule (Kubinyi 1993) which is defined as

$$MR = \frac{n^2 - 1}{n^2 + 2}\left(\frac{M}{d}\right) \tag{3.4}$$

where:
n demonstrates the refractive index
M and *d* are the molecular weight and density, respectively

Another set of important and commonly used whole molecule descriptors is "electronic descriptors". These descriptors are derived from electronic distribution of the molecules that are fundamentally calculated using quantum mechanical theory. There are some examples of these descriptors such as highest occupied molecular orbital (HOMO) and lowest unoccupied molecular orbital (LUMO) energies, dipole moment, polarizability, superdelocalizability, and atomic charges just to mention a few (Karelson et al. 1996, Katritzky et al. 1995, Roy et al. 2015a, Todeschini and Consonni 2010).

HOMO and LUMO are the well-known quantum chemical-based descriptors indicative of molecular reactivity responsible for charge–transfer interactions originated from frontier molecular orbital theory (Karelson et al. 1996). In a molecule with ionization capability, the electrons can be easily transferred; in other words, the nucleophilicity of the molecule is characterized by HOMO energy, whereas the electron affinity of a molecule is subjected to LUMO energy as a measurement of electrophilicity of a molecule (Roy et al. 2015a, Todeschini and Consonni 2010) (Figure 3.4).

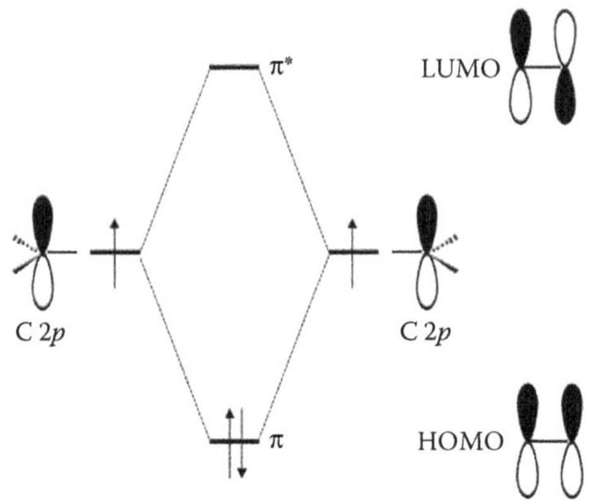

FIGURE 3.4 Schematic representation of HOMO and LUMO energies of molecular orbitals for C=C π bond.

Dipole moment (μ) is a physical vector quantity that provides information regarding global polarity of a compound. This vectorial displacement is directly affected by distribution of positive and negative charges in a molecule (Todeschini and Consonni 2010).

$$\mu = \sum_i q_i . r_i \tag{3.5}$$

where q_i and r_i show charges and positions of the corresponding charges in the molecule and their SI units are coulomb and meter, respectively (i.e., SI unit for dipole moment is coulomb meter known also as Debye) (Figure 3.5).

The other electronic-dependent parameter is polarizability that is dependent on its capability of developing dipole moment when located in an external electrical field.

The last but not the least, parameter of quantum chemical descriptor is the super-delocalizability, which defines the dynamic reactivity of a compound especially in charge–transfer interaction with another molecule considering its contribution in the stabilization energy of molecular interactions (Karelson et al. 1996, Todeschini and Consonni 2010).

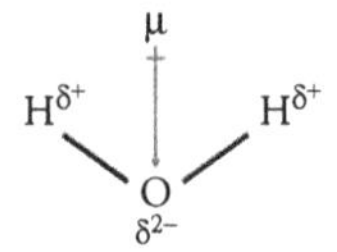

$$\mu = \sum_i q_i . r_i$$

$$\mu = (1.60 \times 10^{-19}\ \text{C}) \times (1.00 \times 10^{-10}\ \text{m}) = 1.60 \times 10^{-29}\ \text{C.m}$$

$$\mu = (1.60 \times 10^{-29}\ \text{C.m}) \times 1\text{D}/(3.336 \times 10^{-30}\ \text{C.m}) = 4.80\ \text{D}$$

FIGURE 3.5 A simple example for calculating dipole moment of water molecule.

3.2.2 Fragment-Based Molecular Descriptors

The other types of molecular parameters employed in QSAR studies are those known as "fragment-based molecular descriptors" that describe a defined part of a molecule. There are different kinds of such descriptors according to their classification based on topology, nodes information of molecular graphs, and their level of fragmentation (Baskin and Varnek 2008a). Using these descriptors is associated with many advantages such as simplicity of calculation and their more easily interpretation of the local effect of structural properties in relation to the activity (Baskin and Varnek 2008b, Zefirov and Palyulin 2002). These descriptors are calculated on the basis of several fragments including atoms, bonds, chains, torsions, common substructures, and basic subgraphs.

The history of fragment-based descriptors dates back to the 1930s in which Hammett (1937) developed an electronic constant (σ_x) for different meta- or para-substituents of benzoic acid derivatives. This parameter denoted by σ shows the electron donating/withdrawing property of a substituent that is expressed as the following formula:

$$\sigma_x = \log\frac{K_X}{K_H} \tag{3.6}$$

In this equation, K_X and K_H indicate the equilibrium constant in substituted and unsubstituted benzoic acid derivatives, respectively.

Another interesting example of fragment-based descriptors is dependent on Free–Wilson analysis in which the presence or absence of molecular fragments is encoded by binary values (i.e., zero and unity) in a congeneric series of molecules. In this method, the physicochemical properties of the substituents are not described directly in the developed QSAR equations and thus are not related to the activity. Instead, it is the presence or absence of the substituents that implicitly define such properties and are correlated to the activity (Craig 1974). However, one should bear in mind that the Free–Wilson and Hansch analyses are fundamentally similar and can be considered as the two sides of the same coin.

$$\text{Log}\frac{1}{C} = \sum a_{ij}X_{ij} + \mu \tag{3.7}$$

where:

a_{ij} is the substituent contribution of X_i at position j

μ is the average biological activity

Table 3.1 demonstrates the application of fragment-based descriptors as indicator variables in the Free–Wilson analysis for a set of *N*,*N*-dimethyl-α-bromophenethylamines derivatives. The absence or presence of different substituents in positions *X* and *Y* are denoted by zero and unity values. Also, the observed and predicted values and residuals are calculated in this analysis (Kubinyi 1988). Other examples of the Free–Wilson approach can be found elsewhere (Alkorta et al. 2008, Hamzeh-Mivehroud et al. 2015, Sciabola et al. 2011, Terada and Nanya 2000).

TABLE 3.1
Indicator Variables Presented as Binary Values for Free–Wilson Analysis on *N,N*-dimethyl-α-bromophenethylamines Derivatives

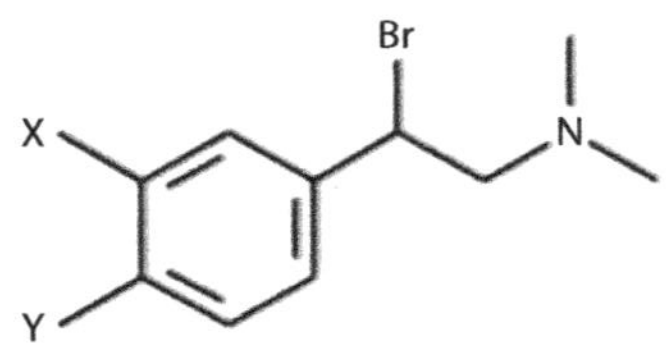

	Structure		Meta-Substituent X						Meta-Substituent X						Log 1/*C* Values		Residuals
No.	X	Y	H	F	Cl	Br	I	Me	H	F	Cl	Br	I	Me	Observed	Calculated	Y_{obs}-Y_{calc}
1	H	F	1	0	0	0	0	0	0	1	0	0	0	0	8.16	8.161	−0.001
2	H	Cl	1	0	0	0	0	0	0	0	1	0	0	0	8.68	8.589	0.091
3	H	Br	1	0	0	0	0	0	0	0	0	1	0	0	8.89	8.841	0.049
4	H	I	1	0	0	0	0	0	0	0	0	0	1	0	9.25	9.250	0.000
5	H	Me	1	0	0	0	0	0	0	0	0	0	0	1	9.30	9.077	0.223
6	F	H	0	1	0	0	0	0	1	0	0	0	0	0	7.52	7.520	0.000
7	Cl	H	0	0	1	0	0	0	1	0	0	0	0	0	8.16	8.028	0.132
8	Br	H	0	0	0	1	0	0	1	0	0	0	0	0	8.30	8.255	0.045
9	I	H	0	0	0	0	1	0	1	0	0	0	0	0	8.40	8.400	0.000
10	Me	H	0	0	0	0	0	1	0	0	0	0	0	0	8.46	8.275	0.185

(Continued)

TABLE 3.1 (*Continued*)
Indicator Variables Presented as Binary Values for Free–Wilson Analysis on *N,N*-dimethyl-α-bromophenethylamines Derivatives

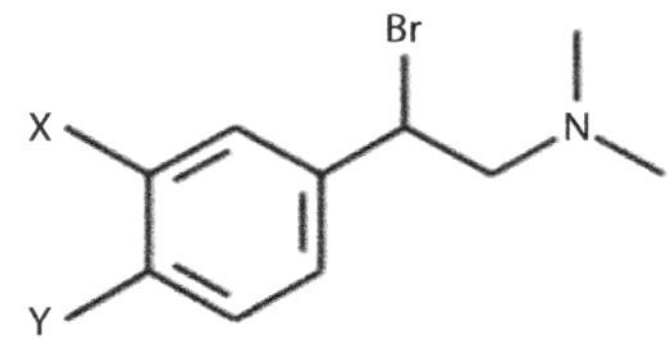

	Structure		Meta-Substituent X						Meta-Substituent X						Log 1/*C* Values		Residuals
No.	X	Y	H	F	Cl	Br	I	Me	H	F	Cl	Br	I	Me	Observed	Calculated	Y_{obs}-Y_{calc}
11	Cl	F	0	0	1	0	0	0	0	1	0	0	0	0	8.19	8.368	−0.178
12	Br	F	0	0	0	1	0	0	0	1	0	0	0	0	8.57	8.595	−0.025
13	Me	F	0	0	0	0	0	1	0	1	0	0	0	0	8.82	8.615	0.205
14	Cl	Cl	0	0	1	0	0	0	0	0	1	0	0	0	8.89	8.796	0.094
15	Br	Cl	0	0	0	1	0	0	0	0	1	0	0	0	8.92	9.023	−0.103
16	Me	Cl	0	0	0	0	0	1	0	0	1	0	0	0	8.96	9.043	−0.083
17	Cl	Br	0	0	1	0	0	0	0	0	0	1	0	0	9.00	9.048	−0.048
18	Br	Br	0	0	0	1	0	0	0	0	0	1	0	0	9.35	9.275	0.075
19	Me	Br	0	0	0	0	0	1	0	0	0	1	0	0	9.22	9.295	−0.075
20	Me	Me	0	0	0	0	0	1	0	0	0	0	0	1	9.30	9.531	−0.231
21	Br	Me	0	0	0	1	0	0	0	0	0	0	0	1	9.52	9.511	0.009
22	H	H	1	0	0	0	0	0	1	0	0	0	0	0	7.46	7.821	−0.361
Sum			6	1	4	5	1	5	6	4	4	4	1	3			

Source: Kubinyi, H., *Quantitative Structure–Activity Relationships*, 7(3), 121–133, 1988.

Another way of using fragment-based descriptors is the application of substructure counts in the form of numerical integer values. In this context, certain fragments are considered and their frequency of occurrences in a data set is counted and used to prepare a matrix of descriptors for QSAR analysis. An illustrative example of such parameters is shown in Table 3.2 for the target data set of three structures.

TABLE 3.2
The Representation of Fragment Descriptors Calculated Based on Different Fragment Occurrences for Compounds

	Fragment Descriptors				
Data Set					
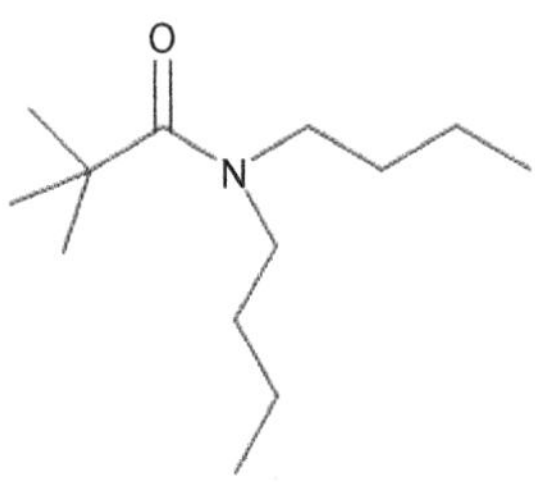	0	10	1	5	0
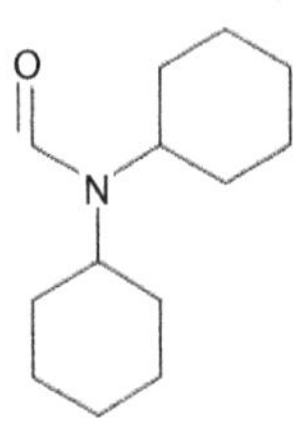	0	8	1	4	0
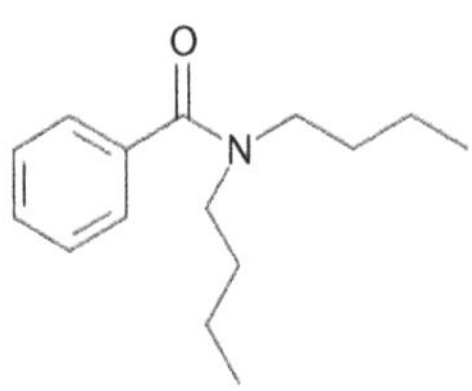	0	4	1	2	4

Source: Baskin, I. and Varnek, A., Fragment descriptors in SAR/QSAR/QSPR studies, molecular similarity analysis and in virtual screening, In *Chemoinformatics Approaches to Virtual Screening*, 1–43, 2008b. Reproduced by permission of The Royal Society of Chemistry.

3.2.3 Dimensionality in Molecular Descriptors

Molecular parameters can be generally classified with respect to their dimensionality into 0-, 1-, 2-, 3-, and 4-dimensional descriptors represented by 0D, 1D, 2D, 3D, and 4D, respectively (Engel 2012, Faulon and Bender 2010, Gozalbes and Pineda-Lucena 2011, Helguera et al. 2008, Lagorce et al. 2011, Roy and Mitra 2012).

Zero-dimensional (0D) descriptors are characterized as the simplest and fundamental descriptors that are derived from chemical formula of the molecules. This type of descriptors is independent of molecular conformation, connectivity, and structure. Constitutional descriptors, counts of atom, and bond type are typical examples of 0D parameters.

One-dimensional (1D) descriptors represent physicochemical properties and counts of molecular groups as fragments and fingerprints that are independent of complete knowledge of structure. Moreover, line notation systems such as simplified molecular input line entry system (SMILES) and Sybyl line notation (SLN) are categorized in this group.

By adding one dimension to the above-mentioned 1D descriptors, 2D descriptors are defined. These conformational independent descriptors are derived from molecular graph theory in which structural topology properties such as molecular connectivity indices, size, branching, and shape are incorporated. The graph theory maps the molecules into a graph in which the atoms and bonds are represented as vertices and edges, respectively. As mentioned before, Weiner index, Randic connectivity index, and Kier–Hall connectivity indices are the examples for 2D descriptors.

Spatial arrangement of atoms results in the generation of 3D descriptors. This type of descriptors corresponds to 3D representation of the molecules and relates to geometrical properties. GETAWAY, WHIM and 3D-MoRSE descriptors, potential energy descriptors, surface area, volume and shape descriptors, and quantum chemical-based parameters are the examples of this type of descriptors, just to mention a few.

Four-dimensional (4D) descriptors represent stereodynamic feature of molecules such as flexibility of bonds and conformational behavior. As an example, molecular interaction field-based descriptors are obtained from calculating the interaction energy between molecule of interest and molecular probes (Cruciani 2005).

3.3 SOFTWARE FOR GENERATION OF MOLECULAR DESCRIPTORS

It is obvious that numerous tools (software) are needed for the calculation of the vast number of molecular descriptors introduced in the field of QSAR analysis. These helpful tools are freely or commercially available for researchers working both in academia or industries. For the sake of convenience, a list of important molecular descriptor generating software is compiled in Table 3.3. In this table, the properties of each software in terms of name, availability, and types of generated descriptors and their website addresses are provided (Danishuddin and Khan 2016, Dong et al. 2015, Helguera et al. 2008, Singla et al. 2013, Todeschini and Consonni 2008a,b, 2010, Vilar et al. 2008).

TABLE 3.3
Important Software for Calculation of Molecular Descriptors Used in QSAR Studies

No.	Software	Descriptor Type	Availability	Web Address
1	ACD/Labs	Physicochemical	Commercial	www.acdlabs.com
2	ADMET predictor	Constitutional, topological, and electronic	Commercial	www.simulations-plus.com
3	ADMEWORKS ModelBuilder	Physicochemical, topological, geometrical, and electronic	Commercial	www.fqs.pl
4	ADRIANA. Code	Constitutional, topological, and electronic	Commercial	www.mn-am.com/products/adrianacode
5	Afgen	Fragment based	Freeware	glaros.dtc.umn.edu/gkhome/afgen/download
6	ALOGPS2.1	Thermodynamic	Freeware	www.vcclab.org/lab/alogps
7	BlueDesc	Constitutional, topological, and geometrical	Freeware	www.ra.cs.uni-tuebingen.de/software/bluedesc
8	CDK	Topological, geometrical, electronic, and constitutional	Freeware	cdk.github.io/cdk
9	ChemAxon	Topological, geometrical, and physicochemical	Commercial	www.chemaxon.com
10	ChemDes	Constitutional, physicochemical, topological, geometrical, and electronic	Freeware	www.scbdd.com/chemdes
11	CODESSA	Constitutional, physicochemical, topological, geometrical, electronic, and thermodynamic	Commercial	www.codessa-pro.com
12	DRAGON	Constitutional, physicochemical, topological, and geometrical	Commercial	www.talete.mi.it
13	E-DRAGON	Constitutional, physicochemical, topological, and geometrical	Freeware	www.vcclab.org/lab/edragon

(Continued)

TABLE 3.3 (*Continued*)
Important Software for Calculation of Molecular Descriptors Used in QSAR Studies

No.	Software	Descriptor Type	Availability	Web Address
14	Filter-it	Constitutional, physicochemical, and topological	Freeware	silicos-it.be.s3-website-eu-west-1.amazonaws.com/software/software
15	ISIDA-fragmentor	Fragment based	Freeware	infochim.u-strasbg.fr/spip.php?rubrique49
16	JOELib	Constitutional, geometrical, and topological	Freeware	www.ra.cs.uni-tuebingen.de/software/joelib
17	MOE	Constitutional, physicochemical, and topological	Commercial	www.chemcomp.com
18	Molconn-Z	Topological	Commercial	www.edusoft-lc.com/molconn/
19	MOLD2	Constitutional and topological	Freeware	www.fda.gov
20	MOLGEN-QSPR	Constitutional, topological, and geometrical	Commercial	www.mathe2.uni-bayreuth.de/molgenqspr/start.html
21	Molinfo	Constitutional and physicochemical	Freeware	chemdb.ics.uci.edu/cgibin/tools/MolInfoWeb.py
22	PaDEL	Constitutional, topological, electronic, and thermodynamic	Freeware	www.yapcwsoft.com/dd/padeldescriptor/
23	PowerMV	Constitutional, topological	Freeware	www.niss.org/research/software/powermv
24	PreADMET	Constitutional, physicochemical, topological, and geometrical	Commercial	preadmet.bmdrc.kr
25	QuBiLS-MIDAS	Constitutional, physicochemical, geometrical, and electronic	Freeware	tomocomd.com/?page_id=238
26	QSARPro	Constitutional, physicochemical, and topological	Commercial	comp.chem.nottingham.ac.uk/download/tmacc

3.4 LIMITATIONS AND CHALLENGES

Various limitations and errors may happen during molecular descriptor calculation. These are summarized in the second and the fifth principles of OECD guideline that are "an unambiguous algorithm" and "a mechanistic interpretation, if possible", respectively. In spite of generation of a large number of descriptors (usually more than thousands of descriptors) using several software, all of them are not interpretable. Use of incomprehensible descriptors can result in a QSAR model that is not useful for mechanistic interpretation. Mostly, the purpose of QSAR studies is to find the mechanism of the action of the molecules on the corresponding targets. According to the OECD guideline, it is advised to employ comprehensible parameters to interpret the QSAR models more easily. Otherwise, such QSAR models containing the noninterpretable descriptors will be useful just for the prediction. Taking into account that there are different algorithms implemented in various software and experimental methods for determining the molecular descriptors, errors in descriptor values can occasionally occur. For example, the calculated value for a descriptor such as Log *P* may vary whether it is measured experimentally or computationally. In such cases, it is advised that the average of consensus values should be calculated and reported along with their errors (Cherkasov et al. 2014, Dearden et al. 2009).

In general, a good descriptor is the one that satisfies some important characteristics such as structural interpretability, simplicity, not measured experimentally, not trivially related to other descriptors, and structural/size dependency (for more details, see Todeschini and Consonni (2008b)).

4 Descriptor Selection

4.1 INTRODUCTION

A journey of a thousand miles must begin with a single step.

Lao Tzu

"Why do we need descriptor selection?" "Are all the determined descriptors essential for QSAR analyses?" "How are the most important molecular descriptors selected?" "Are any of the selected descriptors interpretable?" In the current chapter, it has been tried to provide answers for these questions as much as possible.

What may lead us to use different algorithms for descriptor selection is the large pool of generated molecular descriptors. "Descriptor selection" (or variable selection) is one of the most attractive approaches in many fields of science and engineering and is also the inevitable part of any QSAR study. The more the number of generated descriptors, the higher the computational cost of selecting few of them to be incorporated in the final models. So, the application of molecular descriptor selection methods has found its real place in chemoinformatics analyses.

Significant progresses have been achieved over the recent decades on different preprocessing as well as postprocessing steps to deal with the descriptor selection issue. The preprocessing steps may include the scaling of descriptors (e.g., normalization), descriptor filtering, and construction of train and test sets. However, the postprocessing steps consist of variable selection as well as rescaling approaches. In the current chapter, the concepts, existing challenges, and available software for descriptor selection will be discussed. The reader is also referred to see the practical example in Chapter 7.

4.2 PREPROCESSING OF MOLECULAR DESCRIPTORS

Managing the molecular descriptors is a critical stage in the modeling procedure. Wherever seemed appropriate, the rescaling of the generated parameters may be performed before descriptor selection step in QSAR analyses. The following sections introduce the approaches employed for the pretreatment of molecular descriptors. But, as the very first measure, descriptors with constant (nearly constant or with low variance) or zero values are always removed from the descriptors set.

4.2.1 Scaling

For scaling the molecular descriptors, two well-known scaling methods were employed namely autoscaling (standard normalization) and range-scaling

(Alexander and Alexander 2010). In the case of autoscaling, the entire values of descriptors should be transformed such that they have a mean of zero and variance of unity according to the following formula:

$$X_{ik}^{n} = \frac{X_{ik} - \mu_k}{\sigma_k} \tag{4.1}$$

$$\mu_k = \frac{1}{M}\sum_{i=1}^{M} X_{ik} \tag{4.2}$$

$$\sigma_k = \sqrt{\frac{\sum_{i=1}^{M}\left(X_{ik} - \mu_k\right)^2}{M - 1}} \tag{4.3}$$

where:

the autoscaled descriptors (i.e., X_{ik}^{n}) are calculated by subtracting the mean of descriptors (i.e., μ_k) from the original descriptors (i.e., X_{ik}) divided by their standard deviation values (i.e., σ_k)

i parameter denotes the compound number that varies from 1 to *M*

k (which varies from 1 to *N*) refers to descriptor types

The range-scaling refers to the normalized values obtained from the following equation:

$$X_{ik}^{n} = \frac{X_{ik} - \min X_k}{\max X_k - \min X_k} \tag{4.4}$$

where the range-scaled descriptors (i.e., X_{ik}^{n}) are obtained by subtracting the minimum value of the descriptors set from the original descriptors divided by subtraction of the maximum and minimum values of the descriptors.

The main reason for scaling of molecular descriptors is the existence of significant differences among different descriptors in terms of their range of values, which leads to difficulties in finding out the weighting effect of each descriptor in the model. In other words, descriptors with smaller values will be dominated by those with large values (Alexander and Alexander 2010, Dearden et al. 2009).

4.2.2 Collinearity

The existence of collinearity among molecular descriptors is challenging and should be thoroughly avoided. This needs careful inspection due to the fact that no extra information are incorporated in the model by simultaneous use of collinear descriptors, which leads to the increased number of descriptors and hence, deterioration of the model statistical parameters (Alexander and Alexander 2010, Cherkasov et al. 2014, Dearden et al. 2009). Although there is no threshold value for defining the lack of collinearity between descriptors, a squared correlation coefficient value less than about 0.8 seems to be acceptable as reported elsewhere (Alexander and Alexander 2010, Cherkasov et al. 2014, Dearden et al. 2009).

4.2.3 Train and Test Sets Division

Another data set pretreatment that is highly recommended is to divide the entire data set into train and test sets. Train set is referred to as the set of compounds that directly contribute in model development, whereas the test set is defined as a nonseen set to the trained model that is used for evaluating the predictive power of a trained QSAR model (Roy et al. 2015b). Test set compounds are used in the external validation step of model building, which will be discussed in more detail in Chapter 6. The most important issue in data set division includes the number and distribution of the compounds between train and test sets. Regarding the number of compounds in the series, the suggested size of test set varies from 15% to 20% of the whole data set (Alexander and Alexander 2010). The representatives of the compounds in terms of endpoint values and molecular descriptors should be included in both train and test sets to end up with uniform distributions in the data sets (Roy et al. 2015b). For this purpose, several algorithms are available, which use different criteria for dividing the data set. The more routine selection methods are random selections carried out on the basis of endpoint values and molecular descriptors (Roy et al. 2015b). Purely random selection uses no rational for dividing the data set into train and test sets (Alexander and Alexander 2010). However, the other two rational selection methods mentioned earlier employ the whole range of activities and/or molecular descriptors as the principle of division. In the former, the selection is performed using biological activities (Y-response) of the compounds to allocate them into the corresponding sets. The same procedure can be applied using molecular descriptors. The above-mentioned rational selection methods use some general algorithms such as *k*-means clustering, Kennard-Stone, sphere exclusion, Kohonen's self-organizing map, statistical molecular design, and extrapolation-oriented test set selection methods (Martin et al. 2012, Roy et al. 2008, 2015b).

K-means clustering: In this method, the compounds are divided into *k* groups in which their similarity is assessed according to the mean values for different dimensions for each group. Then a defined subset of compounds (e.g., 20%) from each cluster is selected to be included in the test set, and the rest constitute the train set (Everitt et al. 2011).

Kennard-Stone: Maximum dissimilarity measurement is an important factor calculated on the data set that is used in parallel with hierarchical clustering. The method tries to select a subset of whole data as the test set. The approach is first carried out by selecting two compounds having the greatest geometrical distance of all pairwise comparisons. The third selected compound is the one that has the farthest distance to two selected compounds. Then, a new selection is performed so that its distance to any members of the previously selected compounds in the test set is the greatest. This procedure is repeated until a desired number of compounds are selected (Kennard and Stone 1969, Saptoro et al. 2012).

Sphere exclusion: The theory behind this method is originated from the dissimilarity criterion in which an imaginary sphere is defined around a desired compound usually starting from a distinctive (e.g., the most active) compound with an adjustable radius determining the number of compounds selected and their diversity. Then, the molecules within the sphere are excluded from the selection, and the representative compound is used in the test set. This will be repeated until all compounds are assigned either to test or train sets (Gobbi and Lee 2003, Hudson et al. 1996, Roy et al. 2015b).

Kohonen's self-organizing map: This algorithm finds the closest distance among compounds information to generate the corresponding map in which similar compounds are classified in the same areas of this map. Hence, this process can be easily applied for dividing a data set into train and test sets (Gasteiger and Zupan 1993, Roy et al. 2015b).

Statistical molecular design: This approach is commonly applied for the selection of train set without considering the response variable (Y). In this method, after molecular descriptor calculation, the principal component analysis is used for selecting the best representative groups based on score plots as the train set (Roy et al. 2015b, Tetko et al. 2001).

Extrapolation-oriented test set selection: In this method, a pair of compounds with the highest Euclidean distance in terms of descriptor space is selected and transferred into test set. This is repeated until the desired test set size is achieved. Szantai-Kis et al. (2003) showed that this method performs well when the model had to make a lot of extrapolations.

4.3 DESCRIPTOR SELECTION

In addition to preprocessing step for refining the molecular descriptors, the need for reducing the large pool of such molecular features seems essential. To this end, several algorithms have been developed to satisfy this need to select the most relevant descriptors for building reliable QSAR models. In the next sections, different descriptor selection methods and commonly used tools will be briefly discussed.

4.3.1 Methods and Algorithms

Two strategies that are employed for the feature selection are based on Wrapper and Filter methods (Danishuddin and Khan 2016, Dudek et al. 2006, Shahlaei 2013). Generally speaking, Wrapper methods select a subset of descriptors based on optimizing a fitness function that is sometimes regarded as objective function (or a classifier) with a linear or nonlinear nature (Guyon and Elisseeff 2003, Shahlaei 2013). In contrast to Wrapper methods that take the advantage of using linear or nonlinear regression models, filtering methods eliminate the insignificant descriptors based on statistical parameters such as low-variance and pairwise parameter correlations (Dudek et al. 2006, Shahlaei 2013). Any of these methodologies have their own advantages and disadvantages in terms of complexity, computational cost, and reproducibility. The Filter-based methods are simple, fast, independent of classifier, and are applicable for high-dimensional data sets, whereas Wrapper methods depend on classifiers and are time-consuming and complex (Danishuddin and Khan 2016).

Hybrid-based methods have also been developed by combining these two methods to simultaneously perform preprocessing (Filter-based approaches) followed by Wrapper-based methodologies for descriptor selection. The so-called embedded methods are highly dependent on the structure of the classifier (Danishuddin and Khan 2016, Dudek et al. 2006).

4.3.1.1 Filter-Based Methods

These approaches are typically carried out independently without considering any mapping (data transformation) methods. The correlation-based methods, statistical criteria, and information theory-based methods are some examples just to mention a few (Danishuddin and Khan 2016, Dudek et al. 2006). In any sets of descriptors, the existence of intercorrelated parameters is inevitable and needs to be addressed. Such intercorrelations between the descriptors can be identified by means of correlation-based tools such as Pearson's correlation coefficients. This is performed by setting a threshold value to be systematically calculated between pairs of parameters. In the case of statistical criteria, several methods can be utilized for ranking the descriptors to find the most significant ones in terms of within class criterion. Examples of these statistical tests are the *T*-test, Chi squared test, Fisher's test, and Kolmogrov–Smirnov assessment. Mutual information derived from information theory extracts the relevant parameters on the basis of entropy information embedded in the pairs of random descriptors (Dudek et al. 2006).

4.3.1.2 Wrapper-Based Methods

These types of descriptor selection methods solely depend on two mathematical processes. Those include the usage of objective function and optimization search algorithm to select a subset of important parameters to be used in the next step of QSAR model generation (Danishuddin and Khan 2016, Shahlaei 2013). It is worth mentioning that the way of selecting parameters is highly affected by the used classification algorithm. According to the literature, these methodologies are classified as (1) classical methods, (2) artificial intelligence-based methods, and (3) hybrid methods; it will be explained briefly in the coming subsections.

4.3.1.2.1 Classical Methods

Implementations of algorithms in classical methods are directed by adding and removing molecular features via a regression model. Forward selection, backward elimination, stepwise selection, and selection methods based on prediction are typical instances commonly used in the literature (Shahlaei 2013). In the forward selection method, the best first feature is selected in a heuristic search from the whole descriptors based on both reduced error and the highest fitness function. Subsequent features will be incorporated into the selected subset of the features until the permitted number of descriptors is reached. On the other hand, in the backward elimination, the procedure is initiated with all descriptors followed by checking each descriptor one by one considering that the highest error criterion results in being omitted from the set. Stepwise selection, extensively used in QSAR studies, is similar to forward selection method; however, it does not inherit the straightforward property as observed in the forward selection method. As it simultaneously determines the significance of the previously entered descriptor and if its contribution to the regression model is not significant enough, it will be discarded from the subset of descriptors. The other classical method developed by Liu and colleagues namely variable selection and modeling method based on the prediction (VSMP) is used for supervised descriptor selection in which two criteria are involved (Liu et al. 2003); "interrelation coefficient between the pairs of descriptors (R_{int}) and correlation coefficient (q^2) obtained using the leave-one-out (LOO) cross-validation technique". The readers are referred to this study for further detail (Liu et al. 2003).

4.3.1.2.2 Artificial Intelligence-Based Methods

By incorporating the molecular descriptors and their corresponding biological activities in different mapping algorithms mostly focused on nonlinear relationships, the significant descriptors can be determined using techniques so-called artificial intelligence-based methods. Some examples of these methods are genetic algorithm (GA), simulated annealing (SA), particle swarm optimization (PSO), ant colony system, random forest (RF), and support vector machines (SVM) (Fröhlich et al. 2004, Shahlaei 2013, Svetnik et al. 2003).

GA mimics Darwinian natural evolution in which the initial population composed of defined number of chromosomes is generated for several optimization problems. Apart from many different applications, GA can also be employed in descriptor selection approaches used by QSAR practitioners. In this algorithm, each chromosome encodes a set of random descriptors followed by iteratively crossover and mutation procedures until the purpose of minimizing the error of prediction denoted by fitness function is satisfied using a defined criterion (Danishuddin and Khan 2016, Dudek et al. 2006, Goodarzi et al. 2013, Shahlaei 2013, Sukumar et al. 2014).

With respect to SA, the process is similar to that of GA in finding the global set of variable solutions from search space in stochastic manner but with different algorithms at stages of generating the initial set and preventing from being trapped in local minima. These algorithms are based on metropolis Monte Carlo and Boltzmann distribution, respectively (Dudek et al. 2006, Shahlaei 2013).

PSO, like GA, takes the advantage of random selection of initial population (i.e., particles that are regarded as sets of molecular descriptors) followed by optimization of the fitness function through simulating the birds flocking for finding food. However, there are no mutation and crossover operators. Moreover, in the PSO method, the fitness function is optimized for fitness values of each particle toward the optimum particle during the process. Compared with GA, this approach is easy to deploy with low computational cost (Kennedy and Eberhart 1994, Meissner et al. 2006, Shahlaei 2013).

The other optimization search technique for descriptor selection is ant colony optimization methodology. This method mimics the behavior of ants in finding foods through the shortest path with high level of accumulated pheromone. Descriptors are regarded as ants, and the pathways are considered prediction errors that are supposed to be a single set of molecular descriptors by decreasing the error function (Shamsipur et al. 2009).

RF is also a stochastic and supervised method that is a combination of multiple decision trees in which the subset of descriptors is initially selected on the basis of bootstrapping technique. Then, the importance of descriptors is calculated on the basis of the defined threshold error averaged over the whole forests that then will be voted for their significant contribution. The forest contains various trees, each indicative of several random descriptors. Finally, the most outstanding variables will be ranked for further analyses (Breiman 2001, Genuer et al. 2010, Sokouti et al. 2015).

The other important artificial intelligence-based method in descriptor selection is SVM in which the input data are classified into a number of groups (generally two) in a supervised manner. In this method, all descriptors along with their corresponding biological activities of chemical compounds are denoted by vectors that are used for classification process based on their Euclidean distance matrix. This classification can be carried out in linear and nonlinear approaches. For selection of molecular descriptors, a recursive feature elimination method is employed, in which some weights are set for each vector to determine the significantly important descriptors on the whole classification performance (Liu 2004, Xue et al. 2004).

4.3.1.3 Hybrid Methods

The above-mentioned molecular parameter selection methods can be fused together for rapid and efficient feature selection. For this purpose, first the Filter-based methods are applied for quickly removing the insignificant descriptors with subsequent accurate Wrapper-based algorithms to have the most important descriptors in hand. As an example of hybrid approach, GA-partial least square (PLS) procedure is extensively used in the literature (Deeb and Goodarzi 2010, Hamzeh-Mivehroud et al. 2014, Mehmood et al. 2012, Sagrado and Cronin 2008). In spite of numerous publications that utilize hybrid methods for choosing the descriptors, there is still opportunity for developing novel and improved versions of hybrid methods.

4.4 SOFTWARE AND TOOLS

As the introduction of the concept of QSAR analyses, significant progress has been achieved in the development of tools for descriptor selection, which provide interactive interfaces suitable for practitioners ranging from beginners to experts. Some feature selection algorithms are available as packages for programmable environments (such as R and MATLAB®); however, there are stand-alone software with customized graphical user interfaces (GUIs). Table 4.1 shows the examples of feature selection tools and software.

TABLE 4.1
Representative Examples of Software and Tools for Descriptor Selection

Software	Description
Weka (Hall et al. 2009)	Weka, Java-based GUI, contains several built-in algorithms for different purposes such as data preprocessing, clustering, descriptor selection, model building, and visualization. The most commonly used feature selection tool in this package is SVMAttributeEval
Orange (Demsar et al. 2013)	This package works in a python scripts-based environment with multifunctional tasks including data preprocessing, clustering, descriptor selection, and model building. The feature selection property is carried out by feature scoring technique
R (Team 2015)	R has a programing environment for which several packages are developed. For feature selection purpose, some R packages are available such as modelSampler (Tanujit 2013), glmulti (Calcagno and Mazancourt 2010), VSURF (Genuer et al. 2015), GALGO (Trevino and Falciani 2006), glmgraph (Chen et al. 2015), bestglm (McLeod and Xu 2014), enpls (Xiao et al. 2016), lars (Trevor and Brad 2013) or glmnet (Jerome et al. 2010) (LASSO), RRF (Deng and Runger 2013), randomForest (Liaw and Wiener 2002), penalizedSVM (Natalia et al. 2009), Boruta (Miron and Witold 2010), and caret (Max 2013). In these mentioned packages, selection of important descriptors is done by utilizing various algorithms
MATLAB (TheMathWorksInc 2016)	This software is originally developed for manipulating matrix-based data sets using programming codes and embedded GUI toolboxes. Moreover, developing new toolboxes for different data mining purposes is possible. Some of these toolboxes specifically developed for feature selection are mRMR (Berrendero et al. 2016), SVM and Kernel Methods (Canu et al. 2005), libPLS (Hongdong et al. 2014), GA-PLS (Leardi 2000), and FEAST (Brown et al. 2012)
RapidMiner (Markus and Ralf 2013)	Some classical-based algorithms are utilized for feature selection such as forward selection and backward elimination
ISIDA/QSPR (http://infochim.u-strasbg.fr/spip.php?rubrique53)	The descriptor selection in this software is included in multilinear regression tools. Moreover, the other applications of this software comprise data transformation, validation, and visualization—just to mention a few
FST3 (Somol et al. 2010)	It is a library developed based on C++ programing language. It applies wrappers, filters, and hybrid methods for descriptor selection

4.5 LIMITATIONS AND CHALLENGES

Following the OECD guideline, careful attention is needed to be considered in feature selection procedure. The main challenging issues in this procedure are the use of collinear descriptors, use of excessive numbers of descriptors in QSAR, lack of descriptor autoscaling, and inadequate train/test set selection (Dearden et al. 2009). The first issue, that is, collinearity, violates the OECD principles including the second (i.e., an unambiguous algorithm), forth (i.e., appropriate measures of goodness of fit, robustness, and predictivity), and fifth rules (i.e., a mechanistic interpretation). Employing large numbers of descriptors in QSAR models is problematic, failing to comply with the fourth rule of OECD guideline because using more than five or six descriptors in final model is not statistically significant regardless of the number of data points as reported in the literature (Dearden et al. 2009). However, finding an appropriate and easy-to-use method for feature selection is challenging as there is no straightforward way to choose a limited number of descriptors for QSAR analysis.

As discussed earlier, autoscaling of molecular descriptors is one of those practices that is highly recommended. The performance of the QSAR model can be dramatically influenced by nonscaled descriptors in terms of dominant effect produced by descriptors having a large numerical range.

While dividing the data set into train and test sets, it is critical that the endpoints in both data sets should be well distributed to cover the range of endpoints. The autoscaling and data set division are the subject of the fourth rule of OECD.

Moreover, feature selection procedure is considered as an non deterministic polynomial time (NP)-complete hard problem when one uses an exhaustive forward selection method with time complexity of $O(2^n)$ and hence, a balance is essential to be set between the accuracy and the time spent for feature selection step. This can be achieved by using well-known evolutionary methods for reducing the number of descriptors (Soto et al. 2009).

5 Model Building

5.1 INTRODUCTION

> Yes, we have to divide up our time like that, between our politics and our equations. But to me our equations are far more important, for politics are only a matter of present concern. A mathematical equation stands forever.
>
> **Albert Einstein**

Developing a mathematical relationship, which maps a set of descriptors to the biological activities of compounds, is termed "Model Building" in QSAR analyses. The most important molecular descriptors selected by various descriptor selection algorithms are employed for the purpose of modeling procedure. A wide variety of QSAR models can be derived either in the form of simple and interpretable or complex noninterpretable models. In the QSAR modeling approaches, the ultimate goal is to build the model on the basis of those most relevant descriptors that have the highest correlation with the endpoint values. In this context, the most favorable model is the one with a minimum number of descriptors capable of describing large proportion of the activity data seen for the studied compounds. The data points used for modeling procedure constitute the "train set". However, the best predictive model is selected according to the predictive power of the model for "test set" in terms of statistical analyses (Alexander and Alexander 2010, Dudek et al. 2006, Leach and Gillet 2007).

In general, the model-building approaches are divided into two categories, namely, linear and nonlinear, which are based on the nature of relationship between the selected parameters and biological activities (Alexander and Alexander 2010).

5.2 LINEAR MODELS

Since the beginning, the linear methods in the QSAR analyses have been utilized extensively. This is mostly because of easy interpretation, simplicity, and reproducibility of the generated models (Roy et al. 2015a). There are several methods for establishing linear relationships; among them, the most routinely used ones are explained in the following subsections.

5.2.1 Multiple Linear Regression

The simplest form of linear regression is called simple linear regression (SLR). It is based on a two-dimensional equation (i.e., $Y = aX + b$) between dependent (Y) and independent (X) variables. The constant values for a and b are slope and intercept, respectively. Multiple linear regression (MLR) is an extension of SLR in which more independent variables are involved in a hyperdimensional space (Scior et al. 2009). The general expression for MLR-derived models is shown as follows:

$$Y = \beta_0 + \beta_1 X_1 + \beta_2 X_2 + \beta_3 X_3 + \cdots + \beta_n X_n \quad (5.1)$$

where:

β_0 is the model constant (i.e., intercept)
Y denotes the dependent variable
$X_1, X_2, \ldots, X_n$ are the independent variables representing molecular parameters with their corresponding coefficients $\beta_1, \beta_2, \ldots, \beta_n$

The importance of molecular descriptors as independent variables in describing the biological activities is determined through the magnitude of these coefficients, if the normalized values of independent variables are used in the regression (Yousefinejad and Hemmateenejad 2015). Positive coefficient for a given parameter indicates its direct effect on the biological activity, whereas the negative coefficient for a structural parameter shows its inverse contribution. Apart from this, "p value" should be less than 0.05 for the coefficients to be significant (Roy et al. 2015a). It has been recommended that the number of compounds should be greater than the number of independent variables, and the minimum ratio of 5:1 (number of compounds to number of independent variables) should be considered on the basis of Topliss ratio (Yousefinejad and Hemmateenejad 2015). As mentioned in the previous chapter, the collinearity of the parameters in a QSAR model should be carefully dealt with. Collinear descriptors will result in inaccurate QSAR model in terms of performance.

In a study, Liu et al. (2006) developed an MLR-based QSAR model to predict the estrogenic effect of environmental chemicals with endocrine-disrupting activity. To this end, a collection of 132 compounds was used along with their corresponding endpoint values expressed as log RBA (i.e., relative binding affinity to estrogen receptors). Molecular descriptors were calculated using DRAGON and CODESSA programs. Subsequently, genetic algorithm and MLR were applied for feature reduction and model building using MOBY DIGS software. The model was evaluated by leave-one-out (LOO) cross-validation (CV) and bootstrap methods. For the external validation of the model, D-optimal experimental design, Kohonen self-organizing map (SOM), and random sampling methods were utilized for splitting the data set into train, and test series, using DOLPHIN and KOALA programs. Moreover, an additional data set was used for validating the built model. The results showed that DRAGON descriptors were more reliable for predicating the biological activity of the compounds. The model parameters are illustrated in Table 5.1.

5.2.2 Partial Least Squares

The application of partial least square (PLS) becomes essential when a large number of multicollinear descriptors are used for QSAR analyses. In the PLS technique, the original descriptors ($X_1, \ldots, X_m$) are transformed into "latent variables (LVs) ($t_1, \ldots, t_n$)" that are linear combinations of independent variables. The LVs not only express the variations of molecular descriptors but also model the biological activities (Y) at the same time (Leach and Gillet 2007). The equations for PLS method are expressed as follows:

$$Y = a_1 t_1 + a_2 t_2 + \cdots + a_n t_n \quad (5.2)$$

TABLE 5.1
The Statistics for the MLR-Based Models Developed for the Prediction of Estrogen Activity of the Endocrine-Disrupting Chemicals by Using Different Data Set Splitting Methods

Model Parameters	R^2	Q^2_{LOO}	Q^2_{BOOT}	RMSE	R^2_{pred}	Q^2_{ext}	$RMSE_{ext}$
D-optimal design	0.843	0.799	0.782	0.664	0.789	0.764	0.985
SOM	0.834	0.786	0.768	0.713	0.807	0.785	0.854
Random	0.844	0.797	0.783	0.722	0.761	0.738	0.859

Source: Liu, H. et al., *Chem. Res. Toxicol.* 19, 1540–1548, 2006.

The LVs are described as

$$t_1 = b_{11}X_1 + b_{12}X_2 + \cdots + b_{1m}X_m \tag{5.3}$$

$$t_2 = b_{21}X_1 + b_{22}X_2 + \cdots + b_{2m}X_m \tag{5.4}$$

$$\vdots$$

$$t_n = b_{n1}X_1 + b_{n2}X_2 + \cdots + b_{nm}X_m \tag{5.5}$$

The LVs are orthogonal to each other. The significance of the LVs is assessed practically using CV methods such as LOO and n-fold CV methodologies (more detailed descriptions on CV methods will be covered in Chapter 6) (Yee and Wei 2012).

The application of PLS method in QSAR studies can be exemplified by the investigation of Basant et al. (2012) on a series of p38 mitogen-activated protein kinase inhibitors. The binding pattern and interaction energy were investigated via molecular dynamics simulation. In this work, four interaction energies (i.e., E_{TOT}, E_{VDW}, E_{EL}, and E_{HB}) were extracted for 24 residues important at the binding pocket (24 residues × 4 energy values). So, a matrix of 45 inhibitors × 96 interaction energy values was correlated to their corresponding response vector (i.e., $pIC_{50} = -\log IC_{50}$) using PLS multivariate regression analysis. A three LVs PLS model was selected with root-mean-square error of prediction of 1.36. Based on the developed model, the most important residue interactions (i.e., independent variables) in binding affinity (pIC_{50}) of the studied inhibitors toward p38 were identified. This provides valuable information for designing new scaffolds useful for the inhibition of p38.

5.2.3 Principal Component Analysis

Similar to PLS, principal component analysis (PCA) is utilized for conditions in which the number of descriptors is too large to be amenable for analysis by the MLR technique. PCA method is used to reduce the number of variables required to describe a system without significant loss of information and to exclude redundant information. It is a projection method that helps one to define a few orthogonal variables (known as PCs) based on linear combinations of the original variables leading to effective variable reduction. This method of dimension reduction is also known as "parsimonious summarization" of the data.

In an excellent example of PCA application in QSAR, Yoo and Shahlaei (2017) investigated the anti-HIV activity of a set of CCR5 antagonists. The CCR5 receptors belonging to G protein-coupled receptor family of integral membrane proteins found in many immune cells such as T-cells and could be targeted by human immunodeficiency virus (HIV). The CCR5 antagonists can act as entry inhibitors of HIV and, hence, are promising agents in treatment of AIDS. In this study, 42 CCR5 antagonists along with their corresponding bioactivities were collected from the literature. After generation of the two-dimensional structures of the compounds in CHEMDRAW ULTRA software, they were subjected to energy minimization using AM1 semiempirical method implemented in HyperChem program. Dragon software was used for the calculation of molecular parameters. Molecular parameters were preprocessed by scaling to have equal weights in the analysis. Three significant PCs were selected for the analysis. The results showed that PC1 contributes 51% of the variance and the most important descriptors are sum of atomic van der Waals volumes (Sv), number of carbon atoms (*n*C), number of acceptor atoms for H-bonds (*n*HAcc), and mean atomic polarizability (Mp). The second PC (i.e., PC2) explains 22.92% of the variance that is mainly contributed by H3D (3D-Harary index) and average molecular weight. The PC3 explains 10.9% of variance and the most important parameters in this PC are *n*N (number of nitrogen atoms), *n*X (number of halogen atoms), and *n*HAcc (number of acceptor atoms for H-bonds). Collectively, *n*X, *n*N, and *n*HAcc with respective absolute loading values of 0.990, 1.088, and 0.935, in three first PCs, were determined as the most important parameters (Yoo and Shahlaei 2017). The model was developed on the basis of three orthogonal variables (PCs 1, 2, and 3) that are the linear projection of various structural parameters. Such model benefits from the reduced number of variables and at the same time retains the interpretability.

5.2.4 Free–Wilson Analysis

The use of Free–Wilson analysis in QSAR studies dates back to 1964. In this method, an indicator variable is defined for each substituent in which its presence or absence is denoted by unity or zero, respectively. MLR is employed to correlate the contribution of each substituent to the biological activity in an additive manner. The following general equation can be generated from the following analysis:

$$Y = \mu + \sum_{ij} \alpha_{ij} R_{ij} \tag{5.6}$$

where α_{ij} is the contribution of substituent R_i in position j. If substituent R_i is in position j, then $R_{ij} = 1$; otherwise, it will be equaled to zero. The constant value of μ refers to the theoretical bioactivity of unsubstituted compound (naked compound). For this type of analysis, three assumptions were considered (Leach and Gillet 2007, Scior et al. 2009): (1) a constant contribution of each substituent to the biological activity,

(2) additive contribution of substituents, and (3) lack of interactions between the substituents. Despite the simplicity of this approach, one of the main disadvantages of this analysis is its limitation to predict the biological activity of the compounds having the substituents not involved in the analysis (Leach and Gillet 2007).

Another similar method, which is the modified version of Free–Wilson analysis was proposed by Fujita and Ban (1971). The main difference of this method compared with the Free–Wilson analysis is the use of reference compound (any compound can be selected as reference compound) for which all descriptor binary values are zero and, hence, the corresponding intercept in the model is biological activity of the reference structure. In other words, Fujita–Ban substituent contribution is obtained using subtraction of substituent contribution of Free–Wilson analysis from the corresponding value of the reference molecule (Todeschini and Consonni 2009).

In a study by Sciabola et al. (2008), Free–Wilson QSAR analysis was carried out on kinase inhibitors. Protein kinases are a group of enzymes that can modulate function of a protein by triggering phosphorylation of specific amino acids such as tyrosine, serine, and threonine and have versatile roles in cell growth and movement, metabolic pathways, membrane transport, and gene transcription just to mention a few. Disregulation of kinase activity can cause several diseases such as cancer. Kinase inhibitors are promising small molecules in treatment of different kinds of cancers. On the other hand, the selectivity of these therapeutic agents is a challenging issue in clinical use. In this example, selectivity profile prediction on a set of 700 kinase inhibitors with diaminopyrimidine and pyrrolopyrazole scaffolds (Figure 5.1) was performed using a panel of 45 in-house kinase selectivity screening (KSS) assays.

The summarized procedure for selectivity profile of kinase inhibitors is shown in Figure 5.2.

Briefly, after fragmentation of the given scaffolds, structural matrices were generated for each compound with corresponding inhibitory activity obtained from KSS assay. For each of the above mentioned core structures, statistical analysis was performed on the basis of 45 protein kinase assays. The results showed that the squared correlation coefficients for train set of diaminopyrimidine series vary from 0.82 to 0.95, whereas these values are in the range of 0.73 to 0.93 for pyrrolopyrazole series. The internal LOO CV of the models showed the R^2 of 0.90 and 0.84 for diaminopyrimidine and pyrrolopyrazole series, respectively. For external validation, 30 compounds including 12 diaminopyrimidine- and 18 pyrrolopyrazole-based derivatives

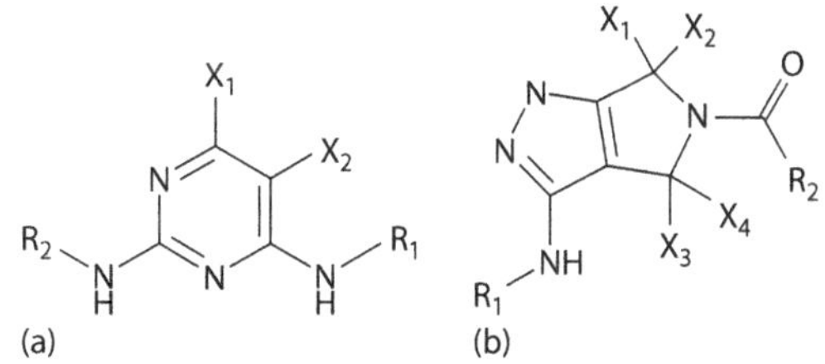

FIGURE 5.1 2D representation of diaminopyrimidine (a) and pyrrolopyrazole (b) scaffolds used for Free–Wilson analysis. X_1 to X_4 are different substituents. R_1 and R_2 are attachment sites for R groups.

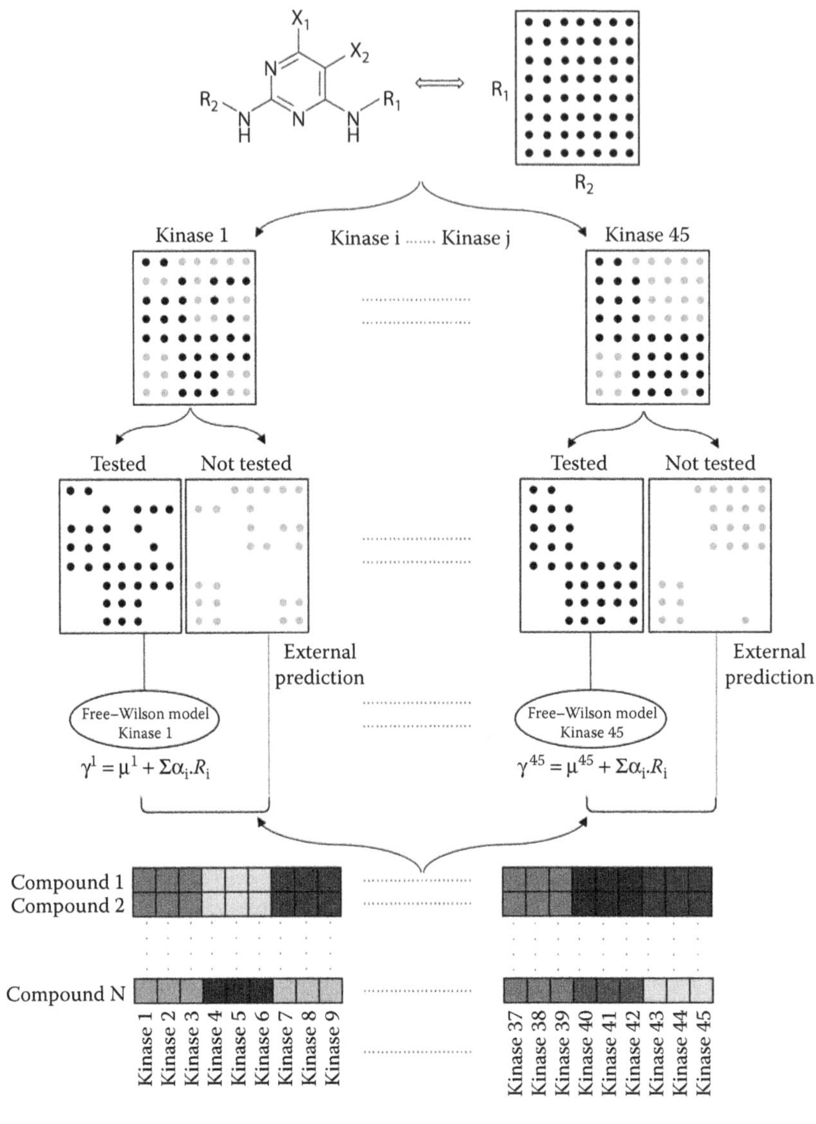

FIGURE 5.2 The Free–Wilson analysis for selectivity profiles in diaminopyrimidine series. A matrix of $R_1 \times R_2$ combination of compounds was generated as R groups were different. They were divided into two series. Black circles were experimentally tested in kinase selectivity screening panel, whereas gray ones were the molecules that were not evaluated. Tested compounds were utilized as train set of Free–Wilson model resulted in the determination of R-group biological activity contribution for every kinase assay. The external validation was performed using "Not Tested" compounds for prediction of kinase selectivity profile. (From Sciabola, S., *J. Chem. Inf. Model.*, 48, 1851–1867, 2008.)

were subjected to the KSS panel. The results revealed that there are agreements between the experimental and the predicted pIC_{50}values with R^2 of 0.77 and 0.78 for diaminopyrimidine and pyrrolopyrazole series, respectively. The investigators were able to determine the selectivity profile on the basis of the contribution of different R groups on kinase inhibitory activity for each model. This provides useful information regarding structural determinants for kinase selectivity (Sciabola et al. 2008).

5.2.5 Linear Discriminant Analysis

Linear discriminant analysis (LDA) works based on classification of the biological activity values and then representing them in class label-based categorical values, but the molecular parameters are in the form of continuous values. The discriminant coefficients or weights (w_i) are determined on the basis of the classification of the studied compounds. A discriminant function is defined for this kind of analysis according to the following equation (Yee and Wei 2012):

$$L = \sum_{i=1}^{k} w_i x_i \tag{5.7}$$

In this expression, L is discriminant score that is linear combination of molecular descriptors $(x_1, x_2, \ldots, x_k)$ and their corresponding discriminant coefficients or weights $(w_1, w_2, \ldots, w_k)$. The discriminant score is used for determining the class of the compounds based on the defined threshold value (cutoff value). The compounds with discriminant scores smaller than cutoff value fall into one group and those with larger values classified into the other group in which a two-group classification is taken into account (Roy et al. 2015a).

Nandy et al. (2013) had developed an LDA model to identify the skin sensitization potential for 147 diverse organic chemicals, which was then used for screening 6,393 compounds derived from the DrugBank database. The chemicals include alkylating agents, sulfonate ester, acrylates, α,β-diketones, and aldehydes. The Marvin Sketch 5.5 software was used to draw the structures of 147 compounds. Then, 161 descriptors were calculated using PaDEL Descriptor 2.11 (22 descriptors) and Dragon 6.0 (139 descriptors) software. The large pool of descriptors was subjected to reduction procedure using a form of molecular spectrum analysis based on the difference values between active and inactive compounds. From 147 compounds, 46 and 101 compounds were identified as skin- and nonskin-sensitizing agents, respectively. Then, based on their structural similarity, the data set was divided into discrimination (27 skin- and 47 nonskin-sensitizing compounds) and test (19 skin sensitizing and 54 nonskin-sensitizing compounds) sets. The LDA along with forward stepwise regression methodology and the receiver operating characteristic (ROC) discrimination approaches were employed using STATISTICA 7.0, SPSS 9.0, and MINITAB 14 software. Several classification validation metrics including accuracy, specificity, sensitivity, precision, F-measure, and many more were used to evaluate the performance of the developed model. The results showed that six descriptors such as quadric index (Q_{index}), number of sulfonates fragments (thio-/dithio) (nSO_3), Dragon branching index (DBI), number of triple bonds (*n*TB), rotatable bond (RBN), and number of nitrogen atoms (*n*N) contributed to discriminant function value represented by ΔP in the LDA model as follows:

$$\Delta P = -10.332 + 45.106 \times \text{DBI} - 11.235 \times n\text{SO}_3 - 13.892 \times n\text{TB} - 8.017 \times Q_{\text{index}} - 21.075 \times \text{RBN} - 0.700 \times n\text{N} \quad (5.8)$$

The compounds with ΔP values greater than 0.5 were assigned as nonskin sensitizing, and those with the ΔP values less than 0.5 were considered skin sensitizing. Moreover, the applicability domain of the above-mentioned model was checked by Euclidean distance algorithm using EUCLIDEAN program, which revealed that the prediction values for two chemicals (Fluorescein-5-isothiocyanate and Kanamycin) were unreliable and regarded as outliers. The authors stated that their model can be used successfully for discriminating skin sensitizers and nonskin sensitizers and hence reducing the animal use, time-consuming, and expensive toxicology screening tests.

5.3 NONLINEAR MODELS

Linear models cannot always be used as a routine approach for QSAR analyses; in some cases, a need for nonlinear methods is inevitable. As its name implies, there is a nonlinear relationship between input (i.e., molecular parameters) and output (i.e., biological responses) values in the data sets. Although these approaches may be known as accurate methods, the generalizability and interpretability are debatable. Moreover, they are more prone to overfitting (Dudek et al. 2006). In this regard, different algorithms have been developed that some of them are briefly described in the following subsections.

5.3.1 Artificial Neural Network

Artificial neural networks (ANNs) are originally inspired from brain function mimicking the functional behavior of neuronal networks. The structure of a feed forward ANN typically consists of three layers, namely, input, hidden, and output layers. These layers have several neurons in which the number of neurons will be affected by the required features as input, complexity of the problem, and the corresponding responses as output. These neurons are linked through weighted interconnections to achieve the predicted response in the output nodes (i.e., neurons). Each ANN model needs to be trained via iterative optimization procedure in which the minimum value of mean squared error (MSE) between the observed and predicted output values is satisfied (Lewis and Wood 2014). Figure 5.3 demonstrates the schematic representation of an ANN structure. In QSAR studies, the input and output neurons are molecular descriptors and biological activities, respectively.

There are numerous examples in which ANN has been extensively used in QSAR studies. For instance, an ANN-based model was developed for predicting the blood–brain barrier (BBB) permeability of the drugs. The BBB is an important biological barrier separating the circulating system from the brain in the central nervous system (CNS). The pharmaceutical agents targeting the receptors in the CNS

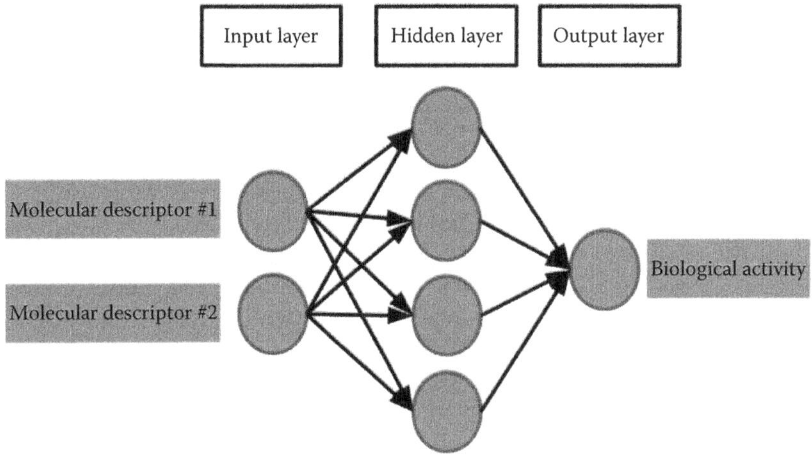

FIGURE 5.3 Schematic representation of an ANN structure.

should cross this biological barrier. The commonly used parameter for showing the permeability of the drug molecules is expressed as brain–plasma ratio (logBB). In the work conducted by Garg and Verma (2006), a data collection of drugs with experimental logBB was used. Seven descriptors comprise molecular weight, topological polar surface area (TPSA), number of H-bond acceptors (HBA), number of H-bond donors (HBD), number of rotatable bonds (NRB), ClogP, and P-glycoprotein (P-gp) substrate probability were calculated using DS ViewerPro Property Calculator, Chem-Office, and ADME Boxes. The data set was divided into train (i.e., 141 compounds) and test sets (i.e., 50 compounds). A nonlinear model using ANN was utilized for the generation of the QSAR model based on a four-layered 7-5-2-1 architecture. Sigmoidal and linear transfer functions were used in hidden and output layers, respectively. Moreover, for weight adjustment feedforward scaled conjugate gradient back-propagation learning algorithm was applied. The result of the generated model is shown in Table 5.2.

Out of 191 compounds, 9 compounds as outliers were excluded from the train set during the modeling procedure. Among the descriptors, molecular weight and number of rotatable bonds showed the highest and lowest impact in predicting the BBB permeability. The developed model was able to show the relationship of P-gp with logBB by demonstrating that 18 molecules were P-gp substrates (probability value of $\geq$0.5).

TABLE 5.2
The Statistics of ANN-Based Model

Parameter	Train Set	Test Set	Working Set
Correlation coefficient (r)	0.90	0.89	0.90
Coefficient of determination	0.82	0.80	0.81
Adjusted R^2	0.81	0.80	0.81
Standard error	0.30	0.32	0.30
Observations	132	50	182

5.3.2 *k*-Nearest Neighbor

In this method, various types of distance measurements such as Euclidean distance or Manhattan distance are employed between the molecular descriptors spaces of a pair of training compounds for classifying them into k groups (k is a predefined value). As there are no training procedures, it is regarded as a "lazy learner". After the k-Nearest Neighbor (k-NN) model is constructed, a test compound can be used for finding the most similar compound in the labeled train set based on its distance from the other existing compounds. However, this method to some extent is prone to misclassification if the k is set too small or too large (Yee and Wei 2012). Moreover, because of its distance-based nature, molecular parameters are needed to be normalized to avoid possible biases in the k-NN model induced by descriptors having greater values compared with other descriptors (Lewis and Wood 2014, Yee and Wei 2012).

A k-NN-based study of a set of 80 3-Arylisoquinolines was investigated against human lung tumor cell line for determining their cytotoxicity effects obtained by the National Cancer Institute protocol based on sulforhodamine B (Tropsha et al. 2011). In this study, seven compounds were selected for external validation procedure and the remaining ones were divided by implementing the sampling sphere exclusion algorithm into six train and test sets differently in terms of the number of included compounds in either train (56 to 61 compounds) or test (12 to 17 compounds)sets. The MolconnZ4.05 software was used for calculating the molecular topology descriptors which were then normalized between zero and unity (i.e., [0, 1]) to avoid unequal weighting effects of descriptors on the final developed QSAR model. Then, the k-NN model optimization was carried out using simulated annealing for stochastic variable selection by satisfying the LOO CV value to be more than 0.5 for the train set. The selected acceptable models would then be validated based on four validation criteria proposed by Tropsha and Golbraikh using the test set which were externally validated by an independent set of (4 to 7) compounds within the applicability domain. The statistics results demonstrated that 12 best k-NN models could be selected (i.e., q^2 (0.81–0.88) and R^2 (0.70–0.86)) where the number of descriptors range from 10 to 28. The Y-randomization test was also applied to avoid the possibility of chance correlation and overfitting. Out of twelve models, only one showed the best performance on the external validation set of seven compounds within applicability domain ($q^2 = 0.81$, $R^2 = 0.93$, $R_0^2 = 0.91$, and $R_0'^2 = 0.93$). The statistics criteria for the selected model are illustrated in Table 5.3. It is worth mentioning that the applicability domain for a compound to be included in or excluded from the test set is represented as a defined threshold distance value between the test compound and the nearest compound in the train set.

TABLE 5.3
The Best *k*NN Model along with the Statistics on Train and Test Sets

Train Set	Test Set	Descriptors	q^2	R^2	k	k'	R_0^2	$R_0'^2$
59	14	14	0.81	0.73	0.99	1.01	0.60	0.73

5.3.3 Decision Tree

One of the methodologies that has a hierarchical-based architecture is known as "decision tree (DT)". The DT procedure is composed of three components: (1) root node, (2) internal nodes, and (3) leaf nodes (Yee and Wei 2012). The starting point of any DT approach is the root node that is denoted by a specific molecular parameter without any incoming branches. The internal nodes rely on defining a conditional statement derived from the remaining molecular descriptors that can be divided into two or more outgoing branches. Furthermore, the final nodes without any outgoing branches are defined as leaf nodes each assigned by a target property (Yee and Wei 2012). Commonly, the DT process can be used both in classification and regression problems. The results obtained from this method will be interpretable when the parameters have been used with comprehensible description (Lewis and Wood 2014). Figure 5.4 illustrates an example of DT for classification of CYP1A2 inhibitors reported by Vasanthanathan et al. (2009).

A combination of DTs predictors constitutes the basis of random forest (RF) algorithm that can be used either for classification or nonlinear regression purposes in cheminformatics (Breiman 2001, Mitchell 2014, Svetnik et al. 2003). The final prediction for the endpoint value by RF models is derived from averaging the outputs of individual DT predictions. RF models are free from overfitting issues that are generally observed for DTs (John 2016).

A practical example for developing QSAR models based on RF algorithm is the study comprehensively carried out on epidermal growth factor receptor (EGFR) inhibitors (Singh et al. 2015). EGFR, a member of receptor tyrosine kinase family, is involved in many different types of cancers and can be a promising target in cancer therapy. In this context, identifying new EGFR inhibitors is of great importance. In this work, an RF-based QSAR analysis was applied for discriminating EGFR

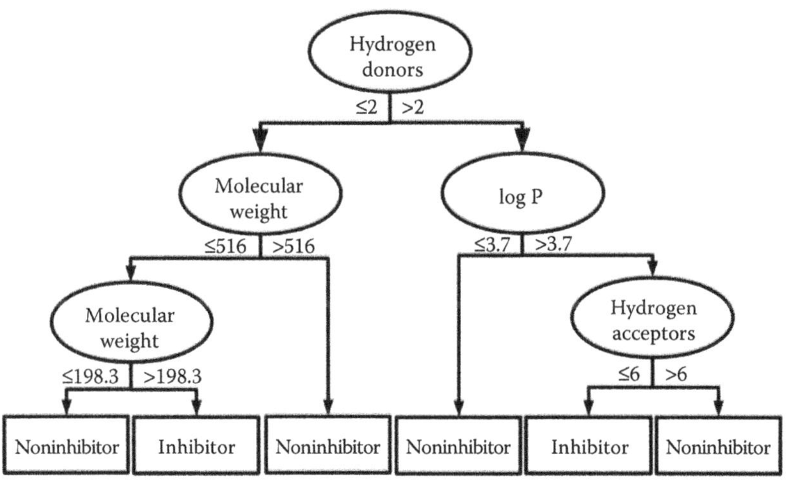

FIGURE 5.4 An example of decision-tree diagram for classification of cytochrome P450 1A2 Inhibitors based on Lipinski's rule-of-five descriptors. (From Vasanthanathan, P. et al., *Drug Metab. Dispos.*, 37, 658–664, 2009.)

TABLE 5.4
The Evaluation Results for Six Classifiers Achieved for EGFR10 Data Set

Classifier	Accuracy	Sensitivity	Specificity	MCC
IBK	82.63	68.69	84.98	0.45
Bayes	70.31	68.73	70.57	0.29
Naïve Bayes	68.23	69.87	67.96	0.27
SVM	83.48	67.11	86.24	0.46
Random Forest	84.95	68.74	87.67	0.49

inhibitors and noninhibitors. The approach was performed on almost 3,500 molecules including quinazoline, pyrimidine, quinoline, and indole scaffolds. In this study, three types of data sets along with their biological activities (i.e., IC_{50}) were constructed from EGFRindb database according to the threshold values of 10 nM (EGFR10), 100 nM (EGFR100), and 1000 nM (EGFR1000) assigning them as inhibitor (<10 nM, <100 nM, and <1000 nM), otherwise noninhibitor. They split the data sets into train and validation sets consisting of 90% and 10% of the data sets. They used PubChem-based 881 binary fingerprints as chemical descriptors for the molecules using PaDEL software. Besides, they took the advantages of frequency of functional groups and maximum common substructures calculations using chemmineR and Chemaxon, respectively, for finding the most active EGFR inhibitors. Then, various classification models were used using fivefold CV to evaluate the performances of different classifiers such as IBK, Bayes, Naïve Bayes, SVM, and RF (available in WEKA software). The results for EGFR10 indicated that the models developed on the basis of RF algorithms using 100 trees had superior prediction performance in comparison with other methodologies (Table 5.4). The evaluation metrics including accuracy, sensitivity, specificity, and Matthews' correlation coefficient (MCC) are shown in Table 5.4. It is worth mentioning that the better results were achieved on the basis of EGFR1000 validation data set with accuracy 85.6%, sensitivity 85.7%, specificity 85.5%, and MCC 0.71. Finally, the authors proposed a user-friendly web server, Python, and R-based packages for identifying anti-EGFR molecules.

5.3.4 Support Vector Machine

Support vector machine (SVM) is a supervised nonlinear approach that can be used for classification and regression purposes (Mitchell 2014). The SVM algorithm generally classifies the data using a hyperplane line by mapping the data into support vectors (i.e., feature space) via kernel functions such as Gaussian, Bernoulli, radial basis, and linear polynomial (Lewis and Wood 2014, Mitchell 2014, Yousefinejad and Hemmateenejad 2015). However, SVMs can also be employed for binary classification and multiclassification problems (Lewis and Wood 2014). The SVM method is known to have less overfitting problem and good generalization for the test sets with higher accuracies compared with other methodologies such as ANN, *k*-NN, and DT (Dudek et al. 2006).

The carcinogenicity is one of the main undesirable features that prevents candidate drugs from being available for clinical use. For this purpose, it is essential to be able to predict the carcinogenicity of chemicals or classify the compounds as carcinogenic and noncarcinogenic before a pharmaceutical agent reaches the market. Zhong et al. (2013) conducted an *in silico* study on 852 noncongeneric chemicals (rat carcinogenicity data including 449 active and 403 inactive compounds) retrieved from the Carcinogenic Potency Database for classifying them as carcinogenics and noncarcinogenics. The Kohonen's self-organizing map (SOM) methodology included in SONNIA software was used for clustering the data set into train and test sets based on MACCS fingerprints (input) and their activity levels (1 or 0 as output). The SOM splitting rules defined in Table 5.5 guaranteed the larger chemical space for train set in comparison with test set.

By applying the above SOM rules, the final train and test sets included 651 (331 active, 320 inactive) and 201 (118 active, 83 inactive) compounds, respectively. Then, the constructed train set was used for both generating SVM classification models as well as 10-fold-CV. Various descriptors including MACCS fingerprints and 334 MOE descriptors (186 2D descriptors and 148 3D descriptors) were used as the compounds' structural representation. The 3D structures of compounds were generated before calculation of molecular operating environment (MOE) descriptors using CORINA software. Twenty four descriptors were selected and normalized in the range of [0.1, 0.9] for modeling by using variance and Pearson correlation analyses (i.e., threshold value 0.9) and stepwise linear regression methodology. Then, the F-Score was employed to evaluate the importance of the selected descriptors in which the higher positive and more negative F-scores were indicative of carcinogenicity and noncarcinogenicity of the compounds, respectively. For performing the SVM analysis, LIBSVM software along with the radial basis function kernel was used with optimum parameters. The metrics for the prediction performance of the best SVM model (A1) are demonstrated in Table 5.6. Moreover, Y-scrambling test used as another model validation method confirmed the lack of chance correlation.

TABLE 5.5
SOM Algorithm Rules for Dividing Data Set into Train and Test Sets

Rules	Number of Compounds with the Same Activity Level	Test Set	Train Set
1	2 or 3	1 out of 2 or 3	1 or 2
2	>3	Half of compounds	Half of compounds
3	1	0	1

TABLE 5.6
Performance Results of the Best SVM Model Developed for Carcinogenicity Prediction

Model	Accuracy (Train Set)	Accuracy (Test Set)	Sensitivity (Test Set)	Specificity (Test Set)	MCC (Test Set)
A1	84.95	80.10	77.11	82.20	0.59

5.4 OUTLIERS

Outliers play a critical role in QSAR modeling, which have been mostly ignored by the researchers working in this field. However, determining "true outliers" is invaluable task requiring more inspection of outliers in terms of their structure and activity. The term "outlier" is defined as any compound that is not well predicted in a QSAR model, in other words any compound that does not fit well in a proposed model (Cronin and Schultz 2003, Verma and Hansch 2005). To categorize the types of outliers, they are divided into structural (so-called leverage) and activity outliers. Structural outliers can be observed because of the fact that different mechanism of action can occur by atypical fragments in molecular structures. The activity outliers originate from "activity cliffs" in the descriptor space in which the properties change dramatically as the result of small structural changes leading to deviation from the hypothesis that "similar compounds have similar biological activities" (Maggiora 2006, Tropsha 2010). To identify and remove the outliers, several methodologies (e.g., statistical techniques, distance based, or SVMs) have been used in the literature in which one or more descriptors space are involved. However, the true removal of the outliers cannot be guaranteed, and hence, it has been suggested that iterative procedures in terms of multiobjective optimization should be employed. Furthermore, in QSAR studies, removal of the outliers is not always a must-to-do task as more inspection is essential for detection of true outliers (Cronin and Schultz 2003, Maggiora 2006).

5.5 SOFTWARE AND TOOLS

Increased interest in application of QSAR studies in different aspects of drug design and development has led to the introduction of variety of algorithms and tools directly or indirectly suitable for QSAR studies. Some of the most commonly used software and packages are listed in Table 5.7.

5.6 LIMITATIONS AND CHALLENGES

Like any other branch of science, the QSAR model building is also not free from challenges. The common sources of the challenging issues are briefly discussed in the following. The most important precaution that can be taken into account to avoid such issues is to detect the compounds with different behavior in terms of their unexpected biological activity and structure in the data set. These compounds that do not fit well in the QSAR model and make the predictive ability of the model questionable are named "outliers" (Verma and Hansch 2005). Peculiar structural variability and unusual functional groups in molecules, experimental errors in determining compounds' biological activity that may arise due to different mechanism of actions for some of the molecules in data set are common sources for rendering a compound an outlier (Cronin and Schultz 2003, Ghandadi et al. 2014). Careful inspection is required for the detection and omission of outliers from the whole data set.

The other critical issue in modeling is "overfitting" of the data. Most of the times, ignoring Topliss and Costello rule for the ratio of train set compounds to molecular

TABLE 5.7
Important Packages and Software Used in QSAR Analyses

Item	Name	Availability	Type	Website
1	DynaFit	Freeware	Nonlinear least-squares regression	https://www.dynafit.com/
2	DataFit/DataFitX	Commercial	Regression analysis (curve fitting)	http://www.curvefitting.com/datafitx.htm http://www.curvefitting.com/datafit.htm
3	FlexPDE	Free for academic use	Package	http://www.pdesolutions.com/
4	Igor Pro	Commercial	Curve fitting	https://www.wavemetrics.com/index.html
5	Minitab	Commercial	Software	http://www.minitab.com/en-us/
6	MLAB	Commercial	Software	http://www.civilized.com/
7	Maple	Commercial	Software	http://www.maplesoft.com/
8	Matlab	Commercial	Software	http://www.mathworks.com/
9	Neural Network add-ins for excel	Commercial	ANN model	http://www.neuroxl.com/
10	NLREG	Commercial	Nonlinear Regression and Curve Fitting	http://www.nlreg.com/index.htm
11	Origin	Commercial	Software	http://www.originlab.com/

(Continued)

TABLE 5.7 (*Continued*)
Important Packages and Software Used in QSAR Analyses

Item	Name	Availability	Type	Website
12	PLS_Toolbox	Commercial	PLS toolbox for Matlab	http://www.eigenvector.com/software/pls_toolbox.htm
13	Weka	Freeware	Software	http://www.cs.waikato.ac.nz/~ml/weka/index.html
14	VEGA	Freeware	Set of packages	http://www.vega-qsar.eu/research.html
15	GA-MLR	Freeware	MLR model	http://teqip.jdvu.ac.in/QSAR_Tools/DTCLab/
16	Orange	Freeware	Software	http://orange.biolab.si/
17	R Project	Freeware	Software	https://www.r-project.org/
18	Octave	Freeware	Software	https://www.gnu.org/software/octave/
19	OECD QSAR toolbox	Freeware	Software	http://www.oecd.org/chemicalsafety/risk-assessment/theoecdqsartoolbox.htm
20	MOE	Commercial	Software	https://www.chemcomp.com/MOE-Cheminformatics_and_QSAR.htm
21	SYBYL-XSuite	Commercial	Software	https://www.certara.com/software/molecular-modeling-and-simulation/sybyl-x-suite/sar-and-qsar/
22	QSARINS	Freeware	Software	http://www.qsar.it/

parameters of 5:1 leads to overfitting of data. This problem is related to the fourth rule of OECD guideline stating "appropriate measures of goodness of fit, robustness, and predictivity". Hence, incorporating of additional descriptors for the purpose of improving fitness of the QSAR model should be carefully managed to avoid the risk of overfitting. Consequently, possibility of chance correlation will be minimized (Cherkasov et al. 2014, Cronin and Schultz 2003, Dearden et al. 2009).

Providing the required statistical measurements when reporting the results for the developed QSAR models is of great value. Models that do not satisfy these requirements will have limited applicability and validity. Inadequate/missing of statistics in a QSAR model is also in relation to the fourth rule of OECD guideline. The QSAR models having good fitness (i.e., correlation coefficient, r^2) should be further evaluated for their predictive power by passing additional statistical criteria known as model validity. Model validation will be fully discussed in the next chapter (Cronin and Schultz 2003, Dearden et al. 2009).

Apart from the above mentioned issues and challenges, a good model should be simple, transparent, and easily interpretable (Cronin and Schultz 2003). Use of ambiguous and noninterpretable descriptors in a QSAR model is associated with several difficulties in terms of interpretation and portability. The more clear the model, the more mechanistically interpretable the biological activity of the studied molecules (Cronin and Schultz 2003). Mechanistic interpretation is crucial, which has been emphasized in the fifth rule of OECD guideline, but practically it is the predictivity power of the model that concerns us the most.

6 Validation of QSAR Models

6.1 INTRODUCTION

QSAR Modeling is not "Push a Button and Find a Correlation".

Gramatica et al. (2012)

Are all the generated QSAR models acceptable? What are the essential criteria for validating the QSAR models? Is the proposed model capable of predicting the biological activities of new chemical entities? These are types of questions that one needs to answer before practically using any QSAR model.

For any QSAR model to be reported as reliable to the scientific community, rigorous considerations should be taken into account. A QSAR model should pass several criteria to be accepted as a good model in terms of predictivity and performance. Despite increasing rate of QSAR-related publications in the literature, only few of them practically satisfy the requirements to be used successfully for prediction purposes. These essential evaluation criteria are generally categorized into internal and external validations (Chirico and Gramatica 2011, Roy et al. 2015b). They are performed on the basis of mathematical equations developed for assessment of the generated QSAR models. As the ultimate goal of developing QSAR models is to minimize as much as possible the errors between the predicted and the observed endpoint values, carrying out several validation measures is a mandatory task for QSAR practitioners.

6.2 VALIDATION METHODS

In the following sections, the most common criteria used for validating the QSAR models will be discussed. These criteria are of great importance for assessing the goodness of the final model. After generating QSAR models, the internal validation procedure is usually performed to select initial QSAR models. Passing the required criteria known as internal validation, the predictive capability of those selected candidate models is evaluated by external validation procedure. Two types of validating approaches are described as follows.

6.2.1 Internal Validation

For internally validating QSAR models, the same data points (compounds) used as train set are employed. All criteria utilized for this purpose are just indicators for the quality of internal assessment for the proposed models. However, the predictive ability of the model may not be generalized for the new compounds.

6.2.1.1 Internal Cross-Validation

In general, the internal cross-validation is done by excluding a small subset of the studied compounds and then performing the model-building procedure by using the rest of the data points followed by predicting the biological activity of the excluded compounds based on the trained model. This is repeated until all compounds are considered as the excluded subset and used for predicting their activity values.

6.2.1.1.1 Leave-One-Out Cross-Validation

In the case of leave-one-out (LOO) cross-validation, which is the most routinely used method for initial evaluation of a QSAR model, one compound is removed from the entire data set followed by training the model using the rest of the studied molecules. The activity of the excluded molecule is predicted by the trained model. In this method, all compounds will be removed one by one until all of the activity values are predicted once. The Q^2 is used for determining the predictive ability of the model in the LOO method.

$$Q^2 = 1 - \frac{\sum \left(Y_{\text{obs(train)}} - Y_{\text{pred(train)}}\right)^2}{\sum \left(Y_{\text{obs(train)}} - \bar{Y}_{\text{(train)}}\right)^2} \tag{6.1}$$

where:

- Y_{obs} and Y_{pred} are the observed and predicted activity values, respectively
- $\bar{Y}$ refers to the average of activity values
- The minimum acceptable Q^2 value is 0.5, and it is obvious from the equation that its value varies from zero to unity

Moreover, standard deviation error of prediction (SDEP) is calculated by the following equation:

$$\text{SDEP} = \sqrt{\frac{\sum \left(Y_{\text{obs(train)}} - Y_{\text{pred(train)}}\right)^2}{n}} \tag{6.2}$$

In this equation, n shows the number of data points.

6.2.1.1.2 Leave-Group-Out Cross-Validation

Another method that is similar to LOO is known as leave-group-out (LGO) (also called as leave-many-out and leave-some-out) in which, instead of excluding only one compound at a time, a defined number of compounds are excluded each time. In this method, the entire data set is divided into k groups, where in every repetition of the process, the biological activities of the excluded group are predicted using the trained model based on the remaining compounds. The corresponding Q^2 is also determined similar to LOO method as mentioned earlier.

6.2.1.2 The r_m^2 Parameter

It should be borne in mind that the high value of Q^2 does not reflect necessarily the good predictive power of the model. Sometimes the high value of Q^2 may be originated from the increased differences in the range of observed values from the mean response value (i.e., $\sum(Y_{\text{obs(train)}} - \overline{Y}_{\text{train}})^2$). In other words, the greater the value of the denominator in Equation 6.1, the greater the value of Q^2. To overcome this issue, a new parameter denoted by r_m^2 was proposed by Roy et al. (2013). The equations for calculating r_m^2 metric are shown below:

$$\overline{r_m^2} = \frac{(r_m^2 + r_m'^2)}{2} \tag{6.3}$$

$$\Delta r_m^2 = \left| r_m^2 - r_m'^2 \right| \tag{6.4}$$

$$r_m^2 = r^2 \times \left(1 - \sqrt{r^2 - r_0^2}\right), \; r_m'^2 = r^2 \times \left(1 - \sqrt{r^2 - r_0'^2}\right) \tag{6.5}$$

$$r_0^2 = 1 - \frac{\sum\left(Y_{\text{obs}} - k \times Y_{\text{pred}}\right)^2}{\sum\left(Y_{\text{obs}} - \overline{Y}_{\text{obs}}\right)^2} \tag{6.6}$$

$$r_0'^2 = 1 - \frac{\sum\left(Y_{\text{pred}} - k' \times Y_{\text{obs}}\right)^2}{\sum\left(Y_{\text{pred}} - \overline{Y}_{\text{pred}}\right)^2} \tag{6.7}$$

$$k = \frac{\sum\left(Y_{\text{obs}} \times Y_{\text{pred}}\right)}{\sum\left(Y_{\text{pred}}\right)^2} \tag{6.8}$$

$$k' = \frac{\sum\left(Y_{\text{obs}} \times Y_{\text{pred}}\right)}{\sum\left(Y_{\text{obs}}\right)^2} \tag{6.9}$$

in the above-mentioned equations, r^2 refers to the squared correlation coefficient between observed and predicted values while intercept is considered. The r_0^2 and r'^2_0 indicate the same concept but forcing the regression line to pass from the origin. For the primed term (i.e., r'^2_0) the X and Y axes are reversed. The slopes of these squared correlation coefficients are k and k', respectively. The recommended values for $\overline{r_m^2}$ (i.e., the average of r_m^2 and r'^2_m) and Δr_m^2 (absolute difference between r_m^2 and r'^2_m) should be more than 0.5 and less than 0.2, respectively.

6.2.1.3 *Y*-Scrambling

Y-scrambling or so-called *Y*-randomization is a useful method for evaluating the robustness of the QSAR model. As the name implies, the model building is performed after scrambling of the endpoint values (i.e., biological responses). A parameter denoted by ${}^{c}R_p^2$ is defined on the basis of the correlation coefficients obtained for the models developed before and after scrambling (i.e., R^2 and R_r^2).

$$ {}^{c}R_p^2 = R\sqrt{R^2 - R_r^2} \tag{6.10} $$

When ${}^{c}R_p^2$ is greater than 0.5, it shows that the original model has not been obtained by chance (Roy et al. 2015b). Expectedly, the randomized models should have lower Q^2 (Tropsha 2003).

6.2.1.4 Bootstrapping

The aim of bootstrapping is to generalize the relationship within the developed model (Clementi and Wold 1995). In this type of validation method, original data set is randomly divided into train and test sets several times. Then, the train set is used in a procedure similar to LOO method with the difference that the exclusion of the compounds is done randomly, and one compound may be excluded once, several times, or never. The division of the original data set into train and test sets is performed few times, and bootstrapping procedure is carried out as described earlier. The obtained averaged values of $Q^2_{\text{bootstrapping}}$ and $R^2_{\text{bootstrapping}}$ should satisfy the minimum acceptable values for Q^2 and R^2 obtained from LOO assessment and should oscillate closely around the Q^2 and R^2 values for the conventional LOO statistics (Kubinyi 1993, Roy et al. 2015b).

6.2.2 External Validation

The other critical "must-to-do" assessment on the generated model is external validation. As mentioned in the previous chapters, before developing a QSAR model, the entire data set is divided into train and test sets based on different algorithms (for comprehensive description, see Section 4.2.3). The test set compounds are not involved in the training of the QSAR model and, hence, are used in external validation procedure. The most recommended criteria for external validation are explained in the following sections.

Selection of a subset of data set as test set is the starting point for external validation. To do this, the biological activities of the test set compounds are predicted for determining the predictive power of the model. The most commonly used parameter for evaluating predictive performance of the model is squared correlation coefficient (R^2) for the test set that is given by following equation:

$$ R^2 = 1 - \frac{\sum (Y_{\text{obs(test)}} - Y_{\text{pred(test)}})^2}{\sum (Y_{\text{obs(test)}} - \bar{Y}_{\text{train}})^2} \tag{6.11} $$

where:

$Y_{\text{obs(test)}}$, $Y_{\text{pred(test)}}$, and $\bar{Y}_{\text{train}}$ are observed, predicted, and average values of biological responses for test and train sets, respectively

R^2 value ranges from 0 to 1, and it is suggested that it should not be less than 0.6

The sole use of R^2 is not sufficient for externally determining the validity of the QSAR models, and additional complementary parameters are usually required for further validation. Some of these criteria, which are extensively used for demonstrating the differences between observed and predicted values of test set activities, were proposed by Golbraikh and Tropsha (2002). In their criteria, two squared correlation coefficients, R_0^2 and $R_0'^2$ are calculated for the correlation established between observed and predicted activities without intercept; however, the latter parameter is calculated on the basis of inverted axes. The slopes of the regression lines are k and k', respectively (Chirico and Gramatica 2012, Martin et al. 2012, Tropsha 2003, 2010). The suggested threshold values for so-called Golbraikh and Tropsha criteria are defined as

$$\frac{\left(R^2 - R_0^2\right)}{R^2} < 0.1 \quad \text{and} \quad 0.9 \leq k \leq 1.1 \tag{6.12}$$

or

$$\frac{\left(R^2 - R_0'^2\right)}{R^2} < 0.1 \quad \text{and} \quad 0.9 \leq k' \leq 1.1 \tag{6.13}$$

$$\left|R_0^2 - R_0'^2\right| < 0.3 \tag{6.14}$$

Discrepancies between observed and predicted biological activity values of the test set in terms of external predictive ability can be calculated through the following equation known as root mean square error of prediction (RMSEP) (Chirico and Gramatica 2011, 2012, Gramatica and Sangion 2016, Roy et al. 2015a,b):

$$\text{RMSEP} = \sqrt{\frac{\sum_{i=1}^{n_{\text{EXT}}} \left(Y_{\text{obs(test)}} - Y_{\text{pred(test)}}\right)^2}{n_{\text{EXT}}}} \tag{6.15}$$

It is worth mentioning that this parameter is not applicable for comparison of the models with different endpoint values due to its dependence on the range of these activity values (Chirico and Gramatica 2011, Gramatica and Sangion 2016).

Another similar criterion for determining the external predictive capability of a QSAR model is the average of absolute errors expressed as (Chirico and Gramatica 2011, 2012)

$$\text{MAE} = \frac{\sum_{i=1}^{n_{\text{EXT}}} \left|\left(Y_{\text{obs(test)}} - Y_{\text{pred(test)}}\right)\right|}{n_{\text{EXT}}} \tag{6.16}$$

The lower the values for RMSEP and MAE, the better the predictivity of the developed model.

In addition to the parameters used for externally validating the models, other criteria have also been suggested by QSAR modelers. These are presented as

Q^2 functions and defined on the basis of Q^2 achieved in LOO internal cross-validation. The first Q^2 function was proposed by Shi et al. (2001) denoted by Q^2_{F1}, which is calculated by using the following formula:

$$Q^2_{F1} = 1 - \frac{\sum_{i=1}^{n_{EXT}} \left(Y_{obs(test)} - Y_{pred(test)}\right)^2}{\sum_{i=1}^{n_{EXT}} \left(Y_{obs(test)} - \bar{Y}_{train}\right)^2} \tag{6.17}$$

Q^2_{F2}, the second Q^2 function was recommended by Schüürmann et al. (2008) with a slight difference in denominator of Q^2_{F1} such that the mean of the observed values in the external set is calculated. This can be regarded as an advantage for conditions that information of the train set is not available.

$$Q^2_{F2} = 1 - \frac{\sum_{i=1}^{n_{EXT}} \left(Y_{obs(test)} - Y_{pred(test)}\right)^2}{\sum_{i=1}^{n_{EXT}} \left(Y_{obs(test)} - \bar{Y}_{test}\right)^2} \tag{6.18}$$

Although both Q^2_{F1} and Q^2_{F2} are utilized by QSAR practitioners, these parameters suffer from nonuniform distribution of train set range. This deficiency has been solved by Consonni et al. (2009, 2010) by developing a new term as Q^2_{F3}:

$$Q^2_{F3} = 1 - \frac{\sum_{i=1}^{n_{EXT}} \left(Y_{obs(test)} - Y_{pred(test)}\right)^2 / n_{EXT}}{\sum_{i=1}^{n_{train}} \left(Y_{obs(train)} - \bar{Y}_{train}\right)^2 / n_{train}} \tag{6.19}$$

This parameter is not dependent on the distribution and size of the external data set (Gramatica and Sangion 2016). However, its value is highly affected by the selection of the train set (Roy et al. 2015b).

The r_m^2 parameter used in internal validation can also be applied similarly in external validation on observed and predicted endpoint values for the test set (Roy et al. 2015b).

A parameter denoted by reproducibility index was also defined by Lin (1989) known as concordance correlation coefficient (CCC). This parameter is used for determining the agreement between observed and predicted endpoint values. The precision and accuracy of the proposed models are measured by this parameter simultaneously. Precision deals with the distance of the observed values from the fitted line, whereas the accuracy detects the deviation of the regression line passing through the origin (Chirico and Gramatica 2011).

$$CCC = \frac{2\sum_{i=1}^{n} \left(Y_{obs(test)} - \bar{Y}_{obs(test)}\right)\left(Y_{pred(test)} - \bar{Y}_{pred(test)}\right)}{\sum_{i=1}^{n} \left(Y_{obs(test)} - \bar{Y}_{obs(test)}\right)^2 + \sum_{i=1}^{n} \left(Y_{pred(test)} - \bar{Y}_{pred(test)}\right)^2 + n\left(Y_{pred(test)} - \bar{Y}_{pred(test)}\right)} \tag{6.20}$$

The ideal value for *CCC* is close to unity; however, the acceptable threshold value is greater than 0.85. Although this threshold is not too rigid, the less scattered the data (observed *vs.* predicted values for the test set), the more accurate the model (Chirico and Gramatica 2011).

All the above-mentioned statistical parameters are critical for assessing the predictive power of the developed models; however, they have their own pros and cons. Gramatica and Sangion (2016) concluded that it is more appropriate to judge on the model predictivity on the basis of Q^2_{F3} and *CCC* statistics that are independent of prediction set distribution and provide more realistic assessment about the model quality, whereas it is too naïve to depend heavily on r^2_m, Q^2_{F1}, and Q^2_{F2} parameters, which are not stable and tend to increase as the size of prediction data set increases (Chirico and Gramatica 2012, Gramatica and Sangion 2016).

6.2.3 Applicability Domain

The ultimate goal of a QSAR analysis is to predict the response values for new chemical entities. In this context, applicability domain (AD) of the train set should be determined. The term AD is defined as "The applicability domain of a (Q)SAR is the physicochemical, structural, or biological space, knowledge, or information on which the train set of the model has been developed, and for which it is applicable to make predictions for new compounds. The AD of a (Q)SAR should be described in terms of the most relevant parameters, i.e. usually those that are descriptors of the model. Ideally the (Q)SAR should only be used to make predictions within that domain by interpolation not extrapolation" (Jaworska et al. 2005).

The splitting of entire data set into train and test sets is of great importance in determining AD. The distribution of molecules in both data sets should cover the whole space in terms of chemical structure and response (Roy et al. 2015). The concept of AD is to avoid incorporating any types of dissimilar compounds in developing the QSAR model. Otherwise, this will result in unreliable prediction of biological activity of new compounds not falling in the range of AD (Golbraikh et al. 2012).

For determining AD for QSAR models, various methodologies have been developed to estimate the interpolation regions in a multivariate space covering descriptors and response values. The common methods are (1) those based on ranges in the descriptor space, or (2) range of the response variable, (3) geometrical methods, (4) distance-based methods, and (5) probability density distribution. For more comprehensive details, readers are referred to Funatsu et al. (2011), Golbraikh et al. (2012), and Roy et al. (2015, 2015b).

An illustrative example of AD is shown in Figure 6.1. As observed in this figure, the plot of standardized residuals versus leverage values of an arbitrary QSAR model is demonstrated. The leverage values are indicative of a compound's distance from the centroid of a given descriptor space. The outliers are determined by applying a squared area in the range of ±3 standard deviations with a leverage threshold of 0.3. These thresholds are arbitrary values for an AD William's plot as demonstrated in Figure 6.1.

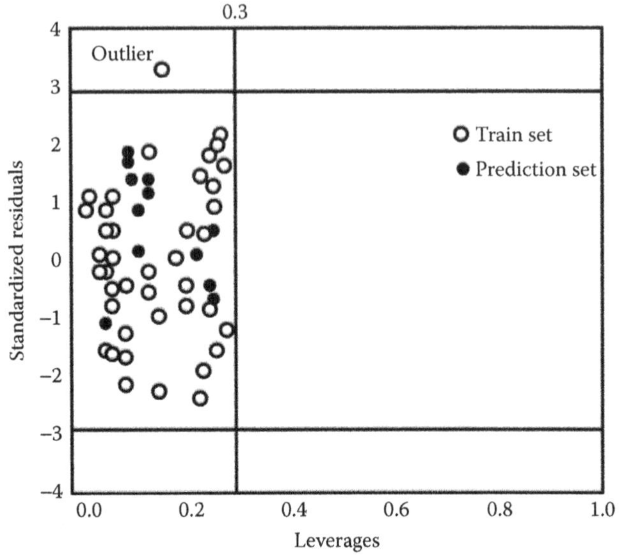

FIGURE 6.1 Applicability domain for outlier determination using William's plot. The outlier is observed via plotting standardized residual versus leverage.

6.3 SOFTWARE AND TOOLS

All the above-mentioned validation methods are implemented in different specialized or general software that are freely or commercially available for QSAR practitioners and modelers. One of the recent software developed for QSAR analysis and validation is QSARINS (Gramatica et al. 2013). In this software, all steps necessary for a QSAR analysis are implemented including data analysis, descriptor selection, model analysis and validation, and model selection. Different validation methods are available including internal cross-validation and external assessments such as RMSE, Q^2_{F1}, Q^2_{F2}, Q^2_{F3}, $\overline{r^2_m}$, Δr^2_m, *CCC*, and Golbraikh and Tropsha's criteria (Gramatica et al. 2013). Roy et al. have developed a set of QSAR validation tools for internal and external validations as well as AD analysis. These have been presented in different programs and are publicly available in http://dtclab.webs.com/software-tools (Ambure et al. 2015). AutoQSAR is an example of commercial software in which different aspects of QSAR analysis are included. One of the advantages of using this package is incorporation of AD analysis as well as validation criteria such as q^2 and RMSE (Dixon et al. 2016). These are few examples of software that can be used as QSAR validation tools, and many more can be found in the literature (Karelson et al. 1999, Katritzky et al. 2006, Todeschini and Consonni 2008b, Vainio and Johnson 2005).

6.4 LIMITATIONS AND CHALLENGES

In spite of existence of rigorous rules for model validation, finding an ideal QSAR model is a troublesome task. Most of QSAR modelers do not pay much attention to OECD guideline, and hence, great numbers of QSAR models are not valid in terms of predictive capability. Even though QSAR analyses are of paramount importance in rational drug-design process, this does not mean that all QSAR models are useful and suitable for predicting the biological activities of new chemical entities. So, careful inspection should be taken into account for evaluating the models based on the above-mentioned validation criteria. However, some researchers may drop a portion of validation criteria to emphasize on the reliability of their developed QSAR model.

One of the limitations in QSAR is the issue concerning with the number of data points for analyses. Using small data set for developing QSAR models prevents performing external validation, and, hence, the reliability of the generated models is limited. For such data, internal cross-validation is performed; however, it will lead to optimistic results as the same data are utilized repeatedly in the process of model-building and assessment (Gramatica 2007, Tichy and Rucki 2009). Having said that, it is worth mentioning that the models suffering from small number of data points may provide valuable information and have applications in their own right. Apart from the number of data points, the division of the data set into train and test sets is a challenging issue as there is a controversy in selecting the data sets. In this context, similar compounds in terms of endpoint values should span both in train and test sets. On the other hand, supervised selection of data sets is not recommended in some sources as it may lead to falsely generating high predictive model (Dearden et al. 2009).

The validation procedure is a critical step in any QSAR analysis in which some of the rules dictated by OECD guideline are covered. According to the third, fourth, and fifth principles of OECD guideline, one should be aware that not following these sets of rules will result in developing unreliable QSAR models (Tichy and Rucki 2009). Inadequate and undefined AD, no consideration of residual plots, inappropriate validation of QSAR models, and lack of a mechanistic interpretation are the main concerns that should be considered in developing a reliable QSAR model (Cherkasov et al. 2014, Dearden et al. 2009).

In conclusion, no QSAR model is considered applicable with high confidence if the required validation criteria are not satisfied.

7 Practical Example

Initiation of a QSAR-based project highly depends on the priorities set by academia and pharmaceutical industry based on unmet worldwide health problem and the availability of the required data. These projects may significantly be of help in drug design and discovery processes by reducing the experimental costs and time.

Now, we are going to demonstrate a practical example of how to perform a QSAR analysis on a real data set. In this example, we are about to analyze the inhibitory activities (expressed as K_i inhibitory constants) of a set of histamine H3 receptor antagonists reported in the literature (Amon et al. 2006, Grassmann et al. 2003, Lazewska et al. 2006, 2008, 2009, Ligneau et al. 2000, Miko et al. 2004, Stark et al. 2001, Wiecek et al. 2011). Compounds having histamine H3 receptor antagonistic activity may be potentially useful in various diseases such as Alzheimer's disease, attention-deficit hyperactivity disorder, epilepsy, narcolepsy, schizophrenia, and obesity (Lazewska et al. 2008).

A search on the PubMed database resulted in nine published research works on nonimidazole- and imidazole-based histamine H3 receptor antagonists reported by Holger's group (Amon et al. 2006, Grassmann et al. 2003, Lazewska et al. 2006, 2008, 2009, Ligneau et al. 2000, Miko et al. 2004, Stark et al. 2001, Wiecek et al. 2011). Table 7.1 illustrates the structures of the compounds collected from these articles in which the biological activities (pK_i) were determined using the same procedure. It is an important consideration to make sure that the activity values collected from different sources are determined by similar procedure and can be combined for the intended QSAR study. The human H3 receptor affinities of the studied compounds were investigated by radioligand binding assay using [125I]-Iodoproxyfan as the competitor. The biological activities (K_i values in nM) range from 0.74 to 1000 nM with ~1350-fold difference between minimum and maximum values. Collectively, 105 compounds were used in the QSAR analysis after excluding 25 compounds due to the reasons listed below. Stereoisomerism is one of the issues that should be dealt with carefully. As biological activity depends on the three-dimensional arrangement of atoms in the space, the configuration of the chiral center clearly has great impact on the activity, and hence, the data for the compounds with mixture of stereoisomers should not be included in the analysis, unless the activities for the pure stereoisomers are known. This is also true for *cis* and *trans* isomers as we do not know how much each isomer contributes to the observed activity. It is obvious if the activity data for different stereoisomers are available, the most active isomer should be included. The biological activities of the compounds should have distinct values and those with ambiguous values such as "less than" or "greater than" (i.e., < or >) should be discarded from the analysis. The quality of the source of data is another important

TABLE 7.1
List of Selected Example Compounds for QSAR-Based Analyses

Compound Number	Structure	Biological Activity	References	Compound Number	Structure	Biological Activity	References
1	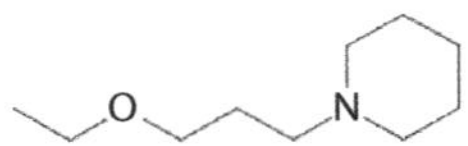	6.15	Lazewska et al. (2006)	54	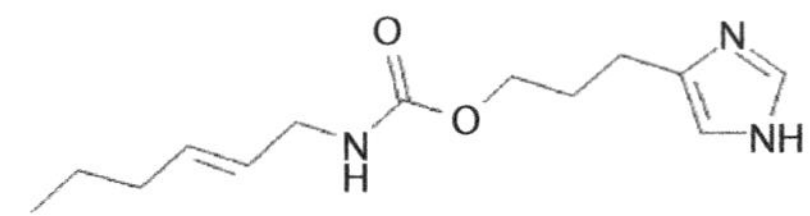	7.62	Wiecek et al. (2011)
2	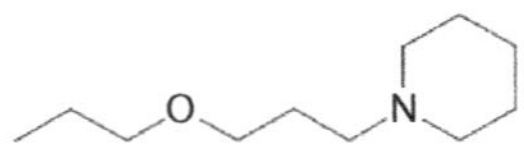	6.13	Lazewska et al. (2006)	55	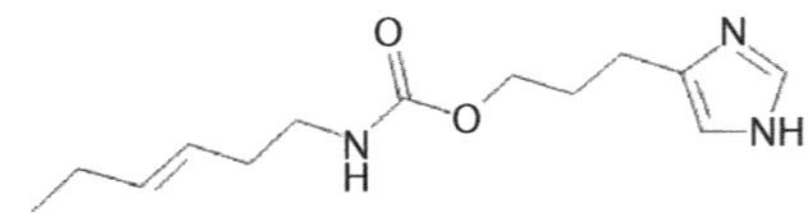	7.59	Wiecek et al. (2011)
3	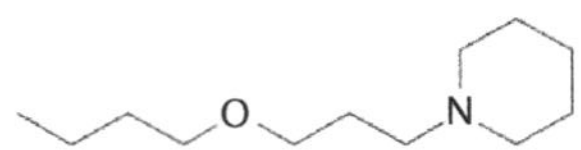	6.14	Lazewska et al. (2006)	56	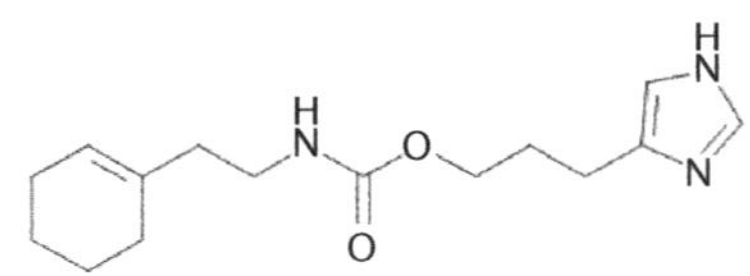	7.59	Wiecek et al. (2011)
4	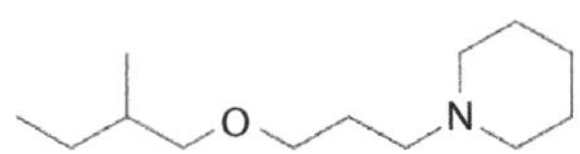	6.15	Lazewska et al. (2006)	57	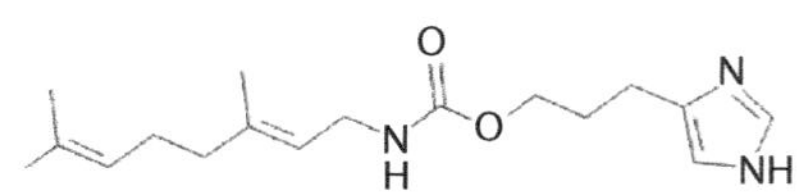	7.7	Wiecek et al. (2011)
5	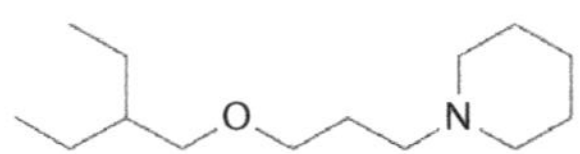	6.14	Lazewska et al. (2006)	58	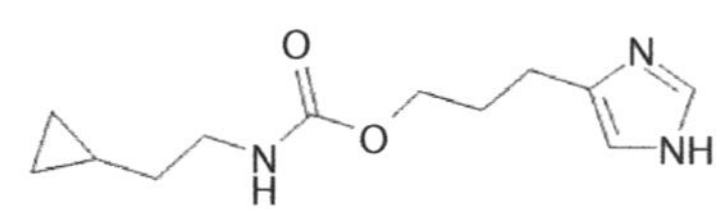	7.77	Wiecek et al. (2011)

(Continued)

TABLE 7.1 (*Continued*)
List of Selected Example Compounds for QSAR-Based Analyses

Compound Number	Structure	Biological Activity	References	Compound Number	Structure	Biological Activity	References
6	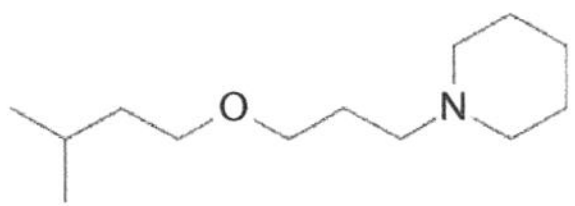	6.25	Lazewska et al. (2006)	59	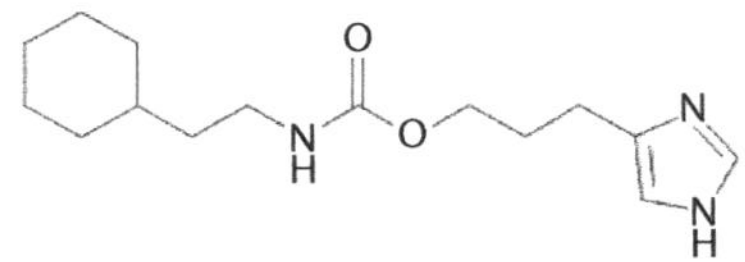	7.6	Wiecek et al. (2011)
7	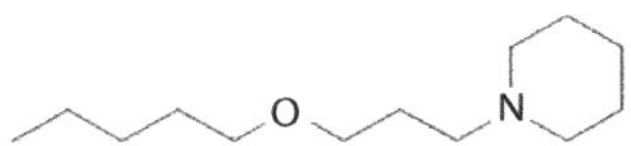	6.47	Lazewska et al. (2006)	60	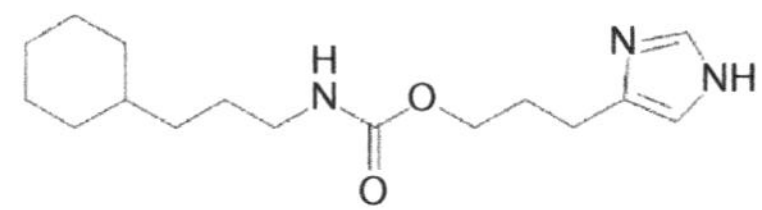	7.07	Wiecek et al. (2011)
8	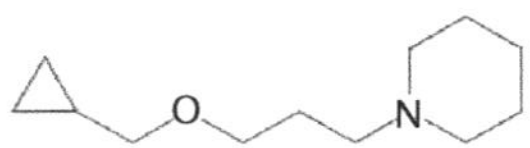	6.47	Lazewska et al. (2006)	61	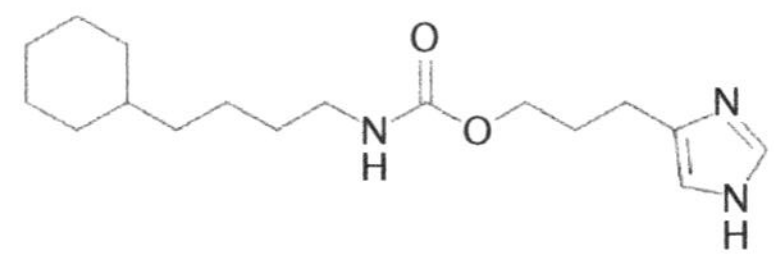	6.71	Wiecek et al. (2011)
9	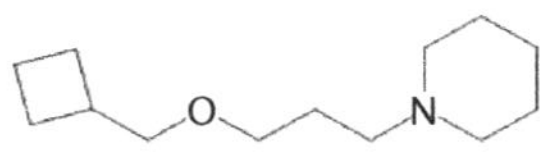	6.51	Lazewska et al. (2006)	62	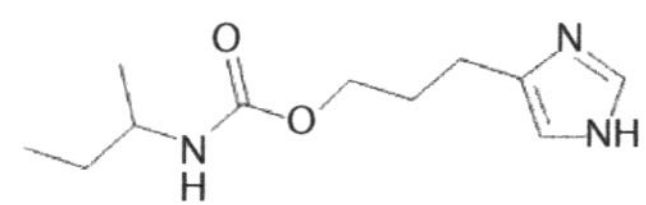	7.92	Lazewska et al. (2009)
10	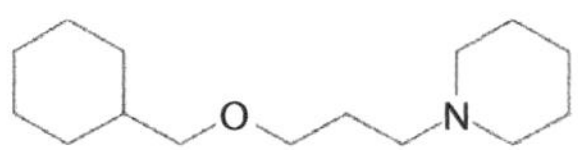	6.42	Lazewska et al. (2006)	63	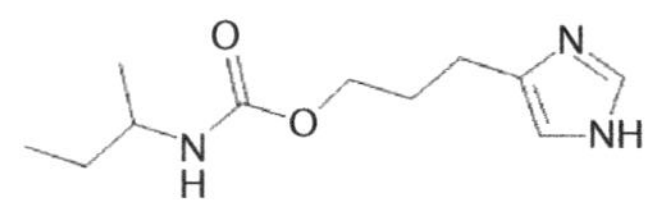	7.51	Lazewska et al. (2009)

(*Continued*)

TABLE 7.1 (*Continued*)
List of Selected Example Compounds for QSAR-Based Analyses

Compound Number	Structure	Biological Activity	References	Compound Number	Structure	Biological Activity	References
11	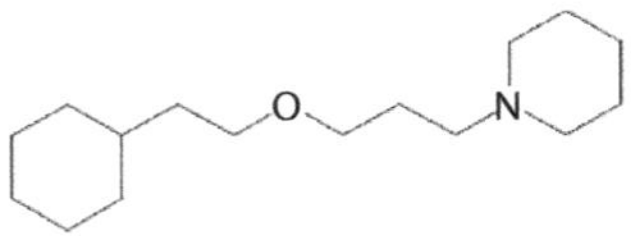	7.40	Lazewska et al. (2006)	64	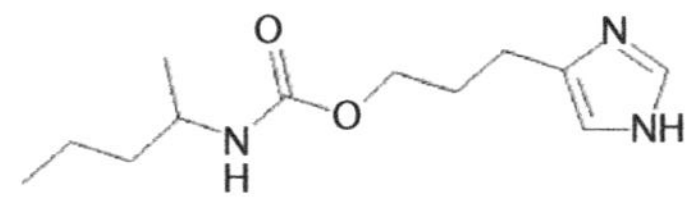	7.82	Lazewska et al. (2009)
12	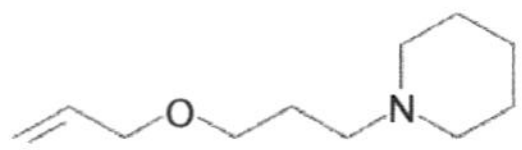	6.41	Lazewska et al. (2006)	65	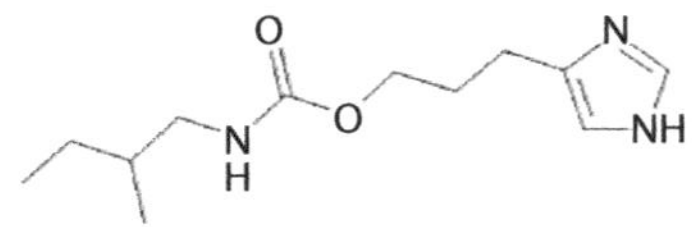	8.33	Lazewska et al. (2009)
13	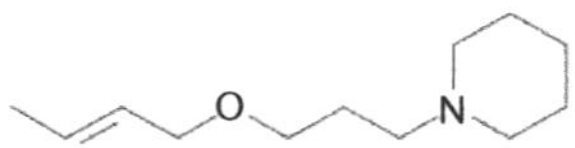	6.28	Lazewska et al. (2006)	66	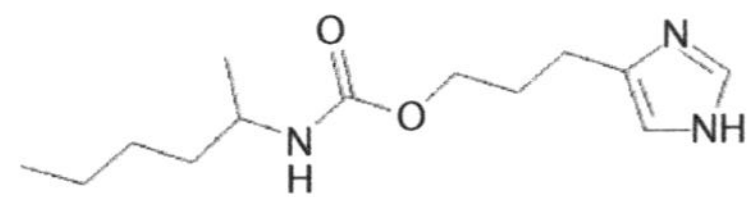	7.72	Lazewska et al. (2009)
14	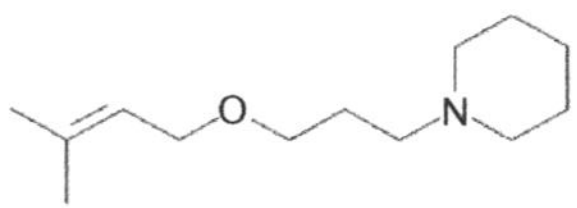	6.47	Lazewska et al. (2006)	67	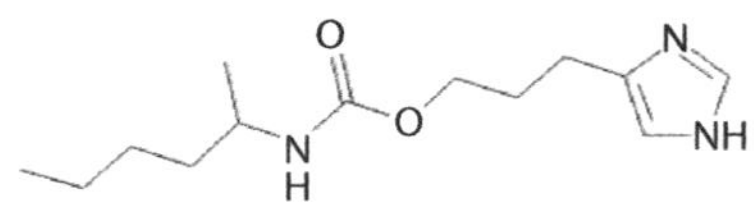	7.38	Lazewska et al. (2009)
15		8.25	Lazewska et al. (2006)	68	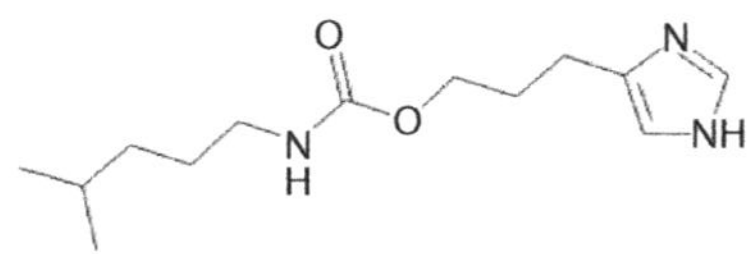	7.89	Lazewska et al. (2009)

(*Continued*)

TABLE 7.1 (*Continued*)
List of Selected Example Compounds for QSAR-Based Analyses

Compound Number	Structure	Biological Activity	References	Compound Number	Structure	Biological Activity	References
16	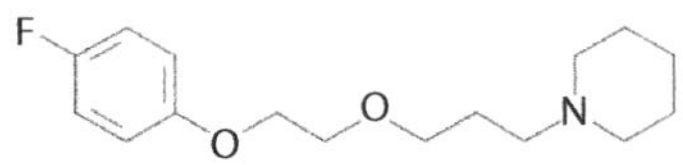	7.51	Lazewska et al. (2006)	69	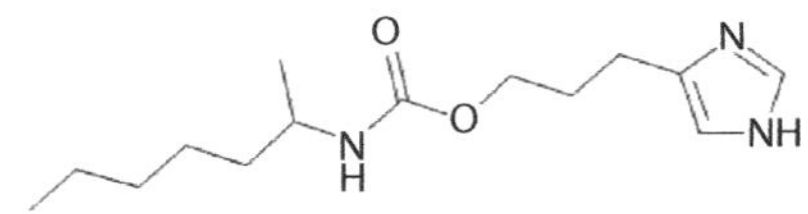	7.39	Lazewska et al. (2009)
17	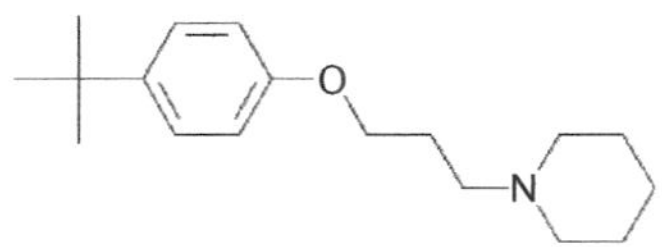	7.66	Lazewska et al. (2006)	70	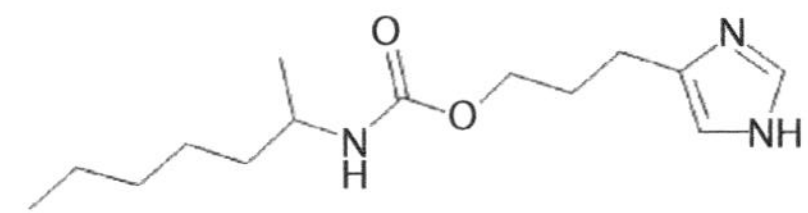	7.13	Lazewska et al. (2009)
18	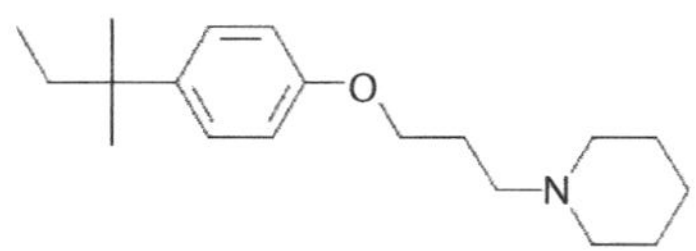	8.08	Lazewska et al. (2006)	71	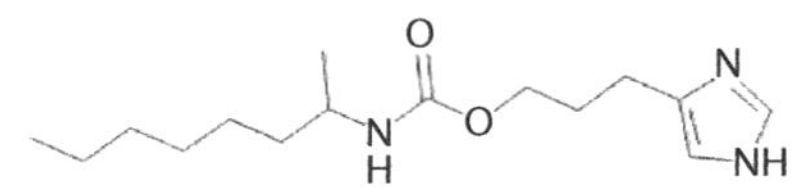	7.55	Lazewska et al. (2009)
19	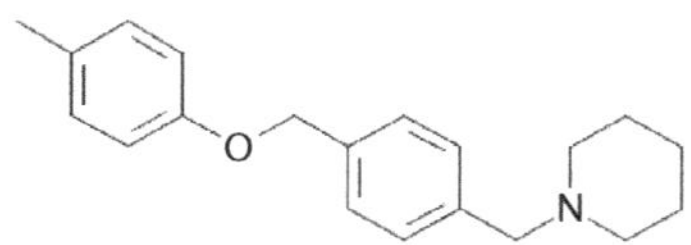	6.9	Miko et al. (2004)	72	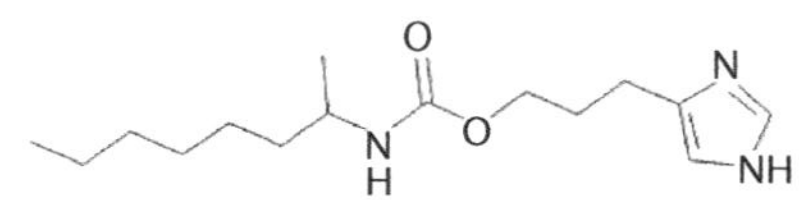	7.37	Lazewska et al. (2009)
20	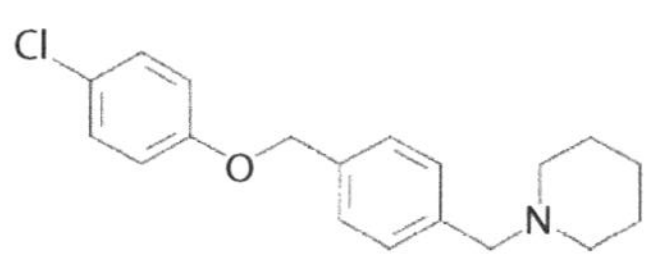	6.9	Miko et al. (2004)	73	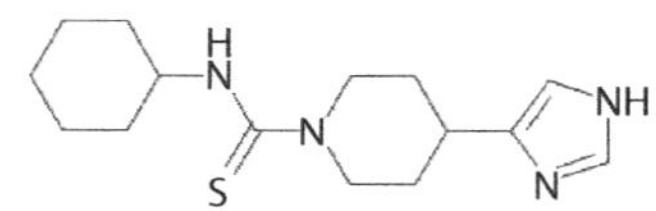	7.22	Lazewska et al. (2009)

(*Continued*)

TABLE 7.1 (*Continued*)
List of Selected Example Compounds for QSAR-Based Analyses

Compound Number	Structure	Biological Activity	References	Compound Number	Structure	Biological Activity	References
21	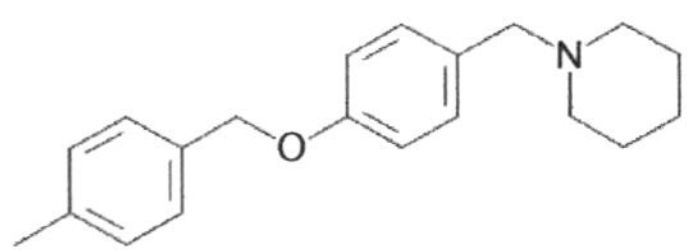	7.4	Miko et al. (2004)	74	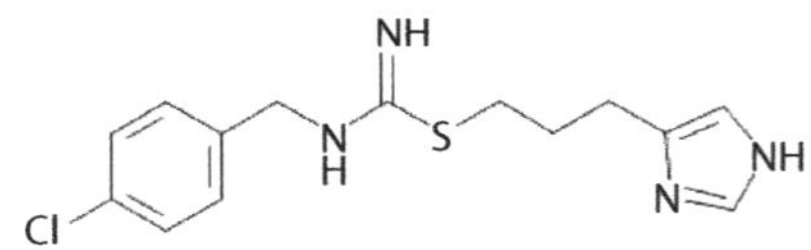	8.62	Lazewska et al. (2009)
22	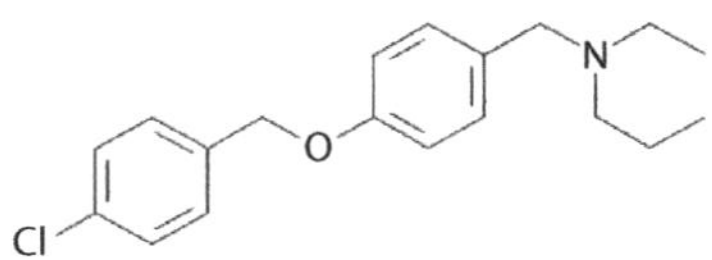	7.1	Miko et al. (2004)	75	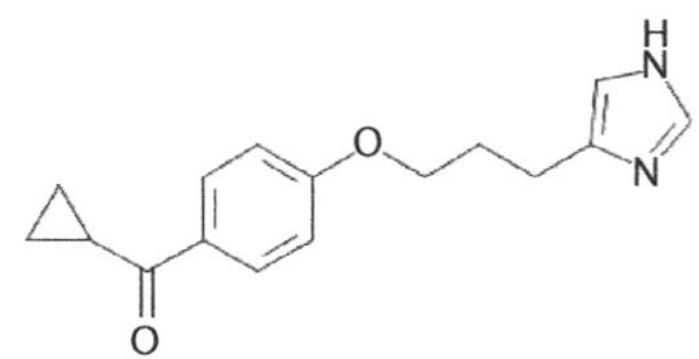	7.34	Lazewska et al. (2009)
23	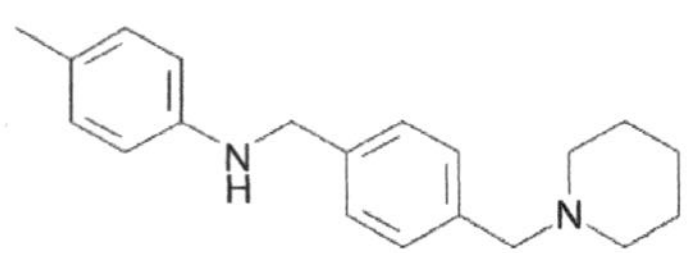	7.1	Miko et al. (2004)	76	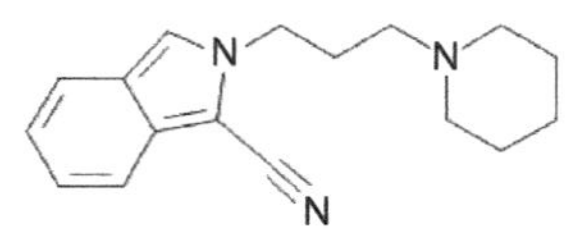	6.45	Amon et al. (2006)
24	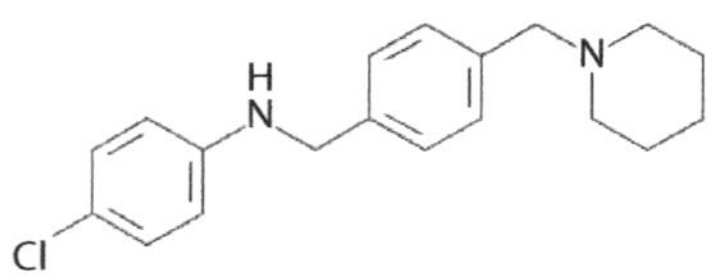	6.9	Miko et al. (2004)	77	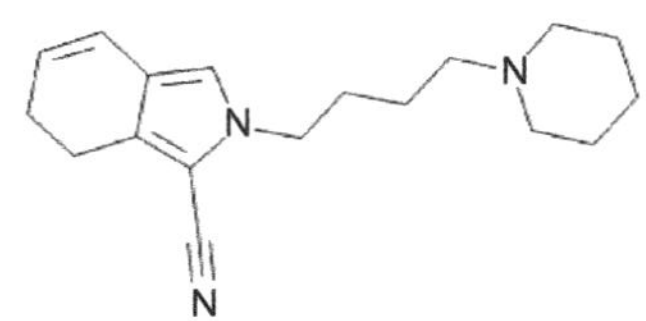	6.75	Amon et al. (2006)

(*Continued*)

TABLE 7.1 (*Continued*)
List of Selected Example Compounds for QSAR-Based Analyses

Compound Number	Structure	Biological Activity	References	Compound Number	Structure	Biological Activity	References
25	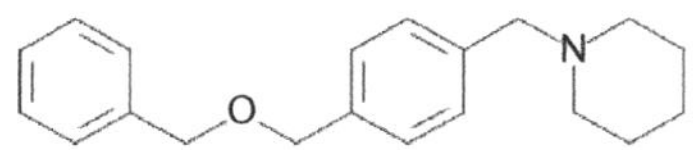	6.7	Miko et al. (2004)	78	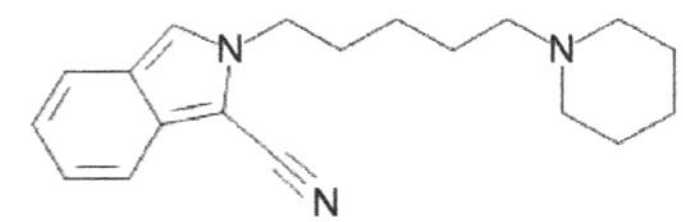	7.51	Amon et al. (2006)
26	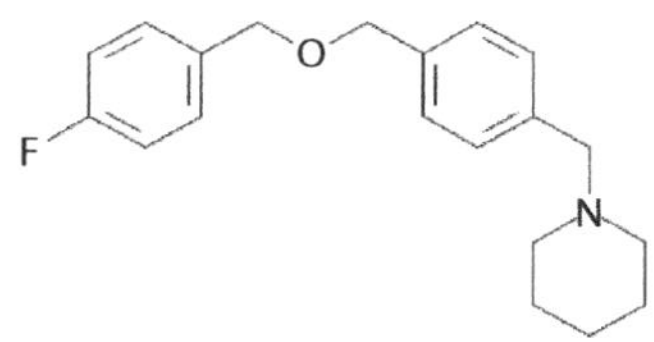	6.6	Miko et al. (2004)	79	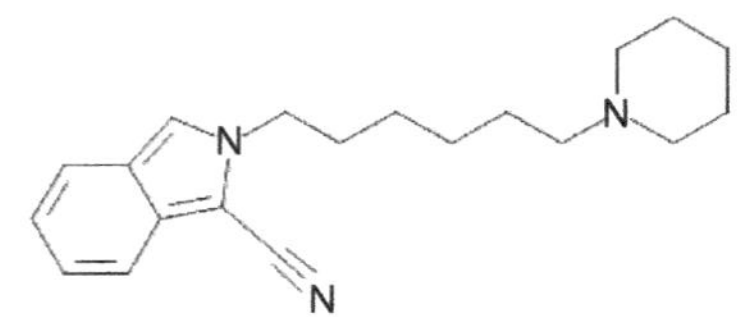	7.51	Amon et al. (2006)
27	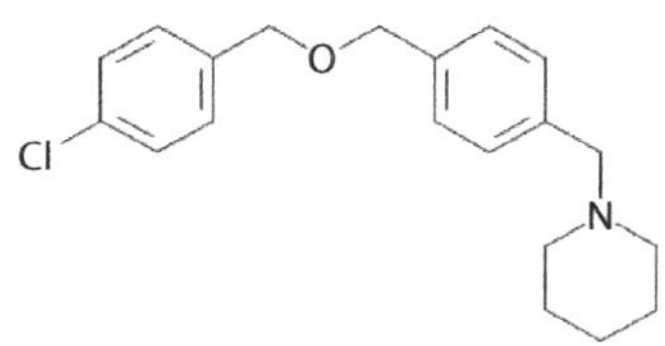	6.8	Miko et al. (2004)	80	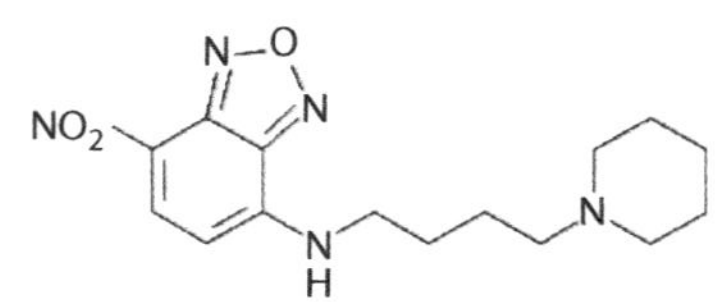	6.77	Amon et al. (2006)
28	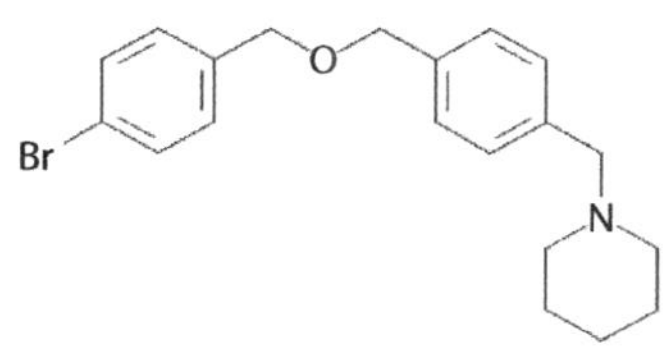	6.7	Miko et al. (2004)	81	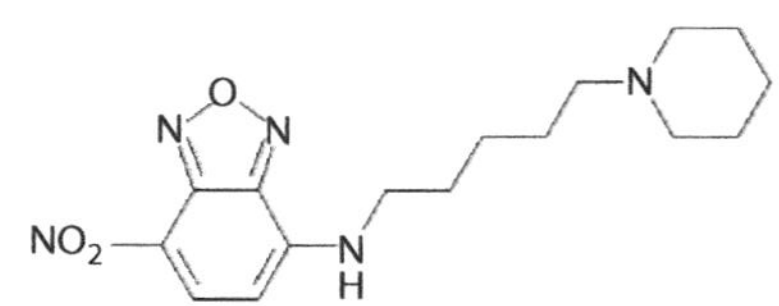	7.96	Amon et al. (2006)

(*Continued*)

TABLE 7.1 (*Continued*)
List of Selected Example Compounds for QSAR-Based Analyses

Compound Number	Structure	Biological Activity	References	Compound Number	Structure	Biological Activity	References
29	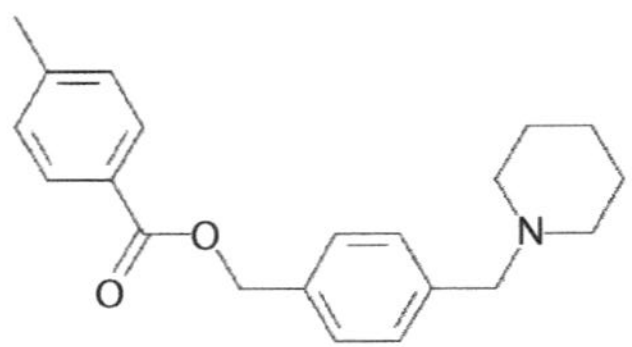	7.4	Miko et al. (2004)	82	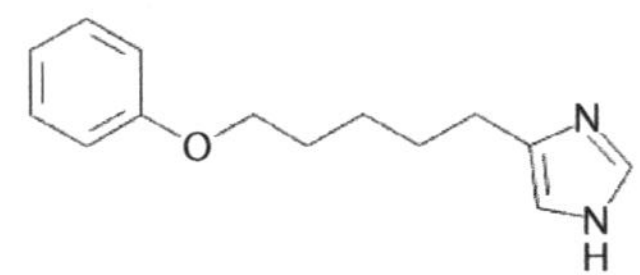	9.13	Grassmann et al. (2003)
30	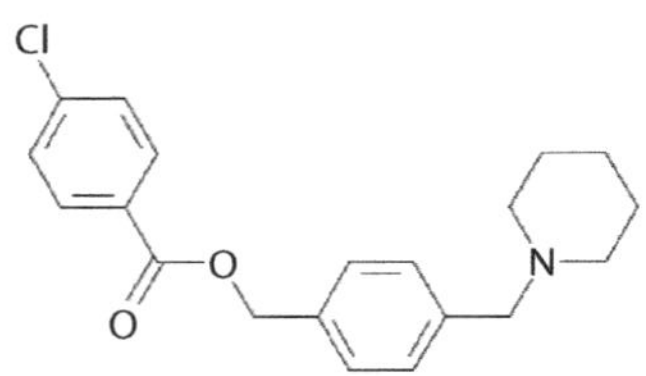	6.6	Miko et al. (2004)	83	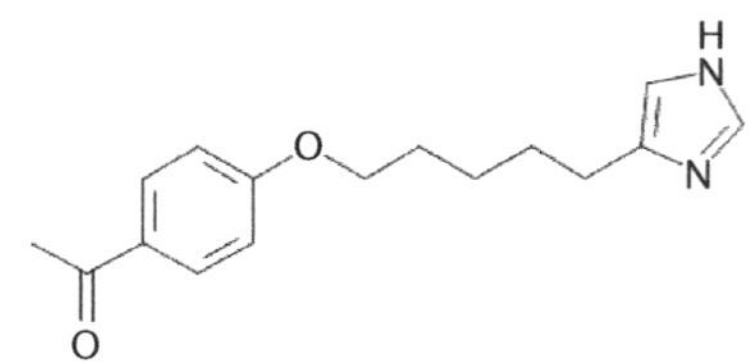	7.35	Grassmann et al. (2003)
31	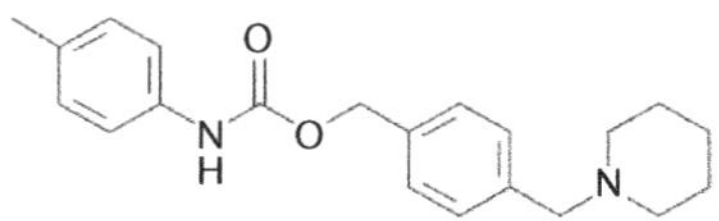	7.1	Miko et al. (2004)	84	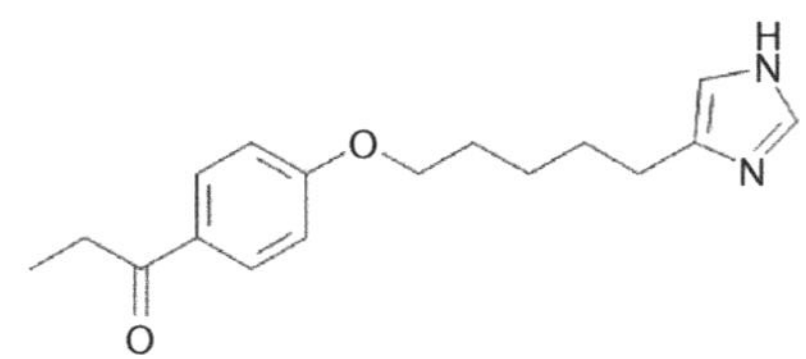	7.07	Grassmann et al. (2003)

(*Continued*)

TABLE 7.1 (*Continued*)
List of Selected Example Compounds for QSAR-Based Analyses

Compound Number	Structure	Biological Activity	References	Compound Number	Structure	Biological Activity	References
32	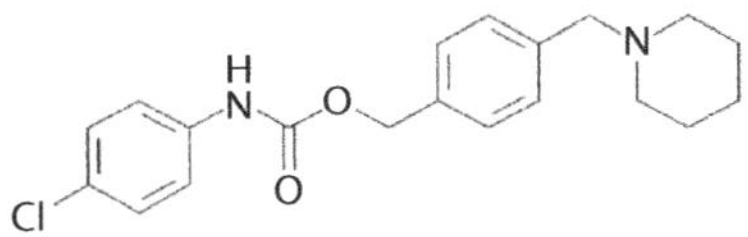	6.7	Miko et al. (2004)	85	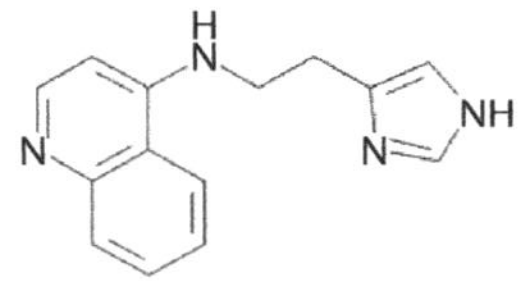	7.7	Grassmann et al. (2003)
33	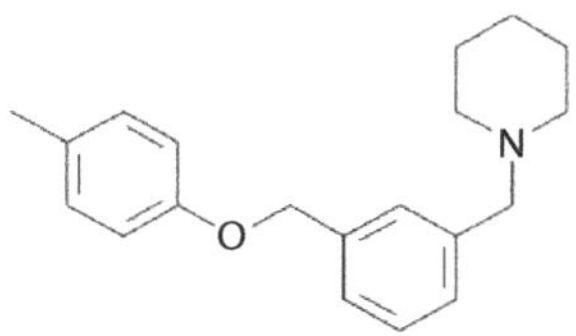	6.3	Miko et al. (2004)	86	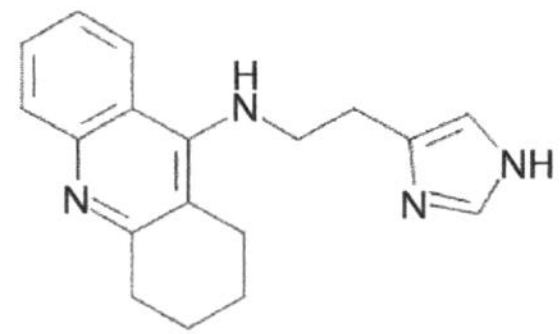	7.68	Grassmann et al. (2003)
34	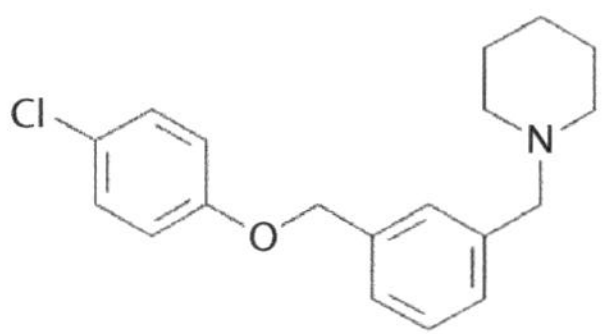	6.6	Miko et al. (2004)	87	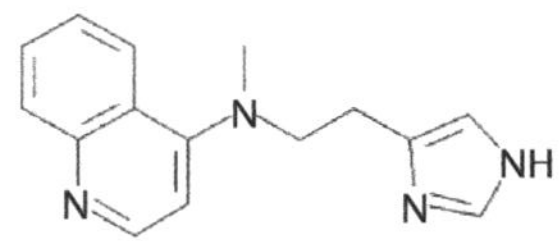	7.33	Grassmann et al. (2003)
35	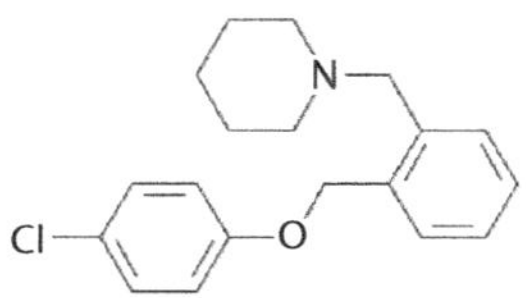	6	Miko et al. (2004)	88	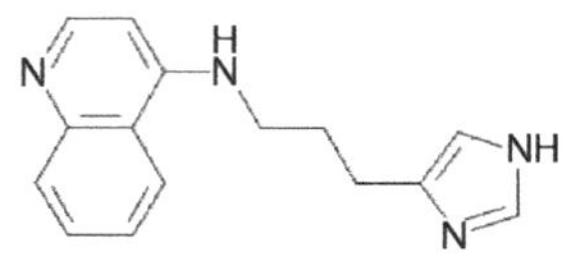	8.39	Grassmann et al. (2003)

(*Continued*)

TABLE 7.1 (*Continued*)
List of Selected Example Compounds for QSAR-Based Analyses

Compound Number	Structure	Biological Activity	References	Compound Number	Structure	Biological Activity	References
36	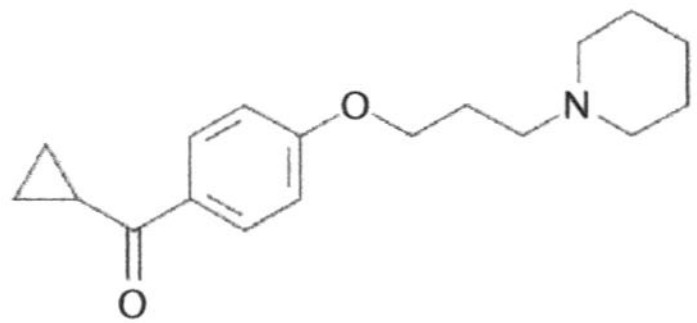	8.8	Miko et al. (2004)	89	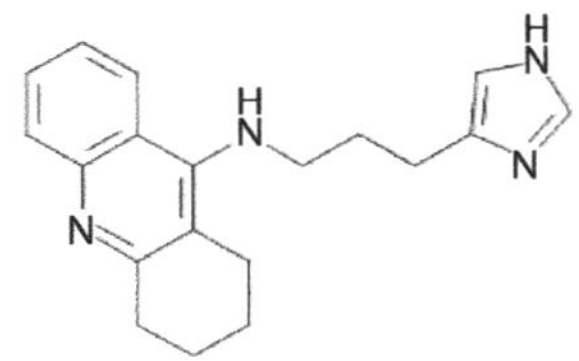	7.03	Grassmann et al. (2003)
37	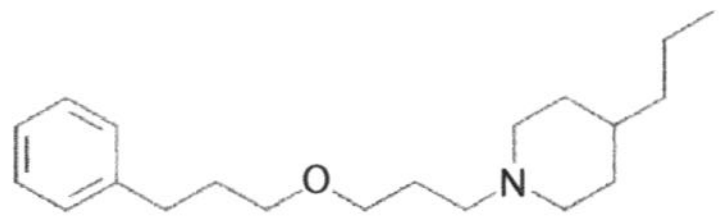	6.6	Lazewska et al. (2008)	90	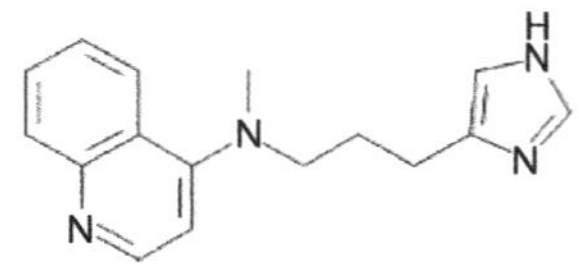	7.77	Grassmann et al. (2003)
38	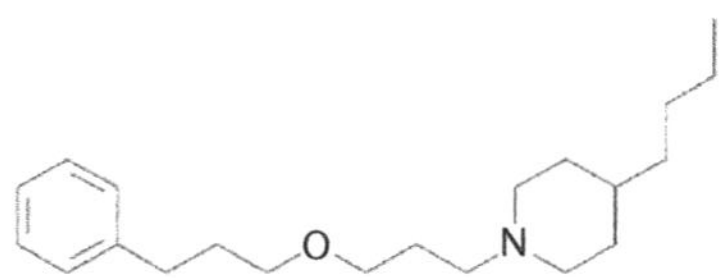	6.56	Lazewska et al. (2008)	91	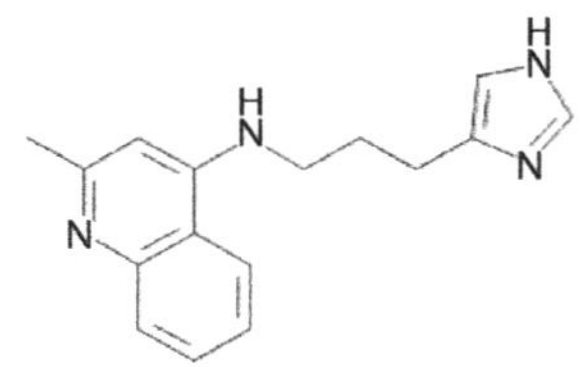	7.29	Grassmann et al. (2003)
39	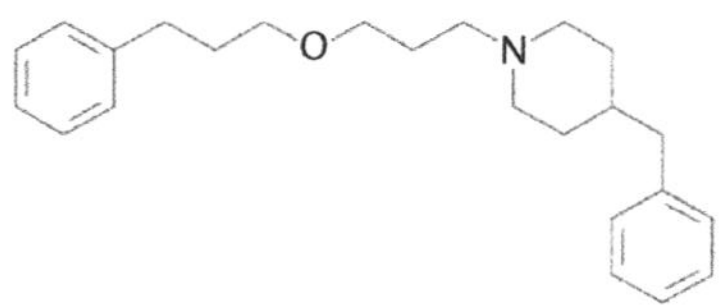	6.49	Lazewska et al. (2008)	92	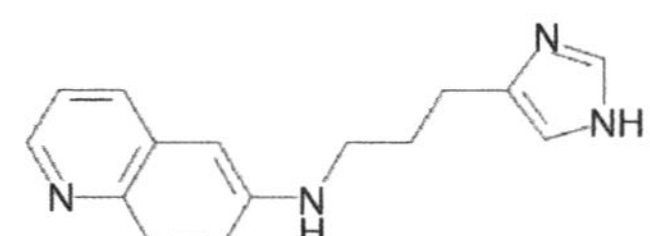	7.15	Grassmann et al. (2003)

(*Continued*)

TABLE 7.1 (*Continued*)
List of Selected Example Compounds for QSAR-Based Analyses

Compound Number	Structure	Biological Activity	References	Compound Number	Structure	Biological Activity	References
40	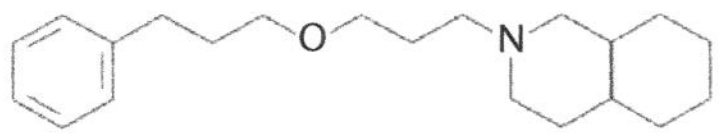	6.41	Lazewska et al. (2008)	93	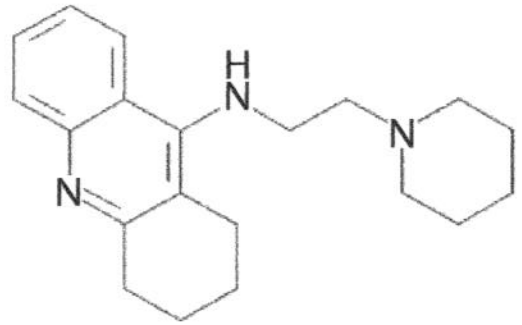	7.72	Grassmann et al. (2003)
41	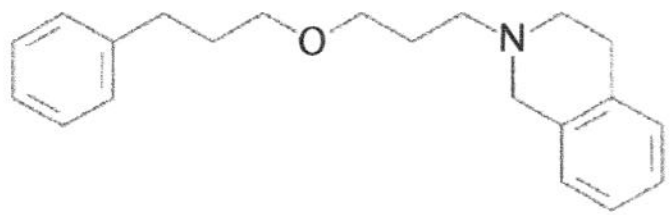	6.25	Lazewska et al. (2008)	94	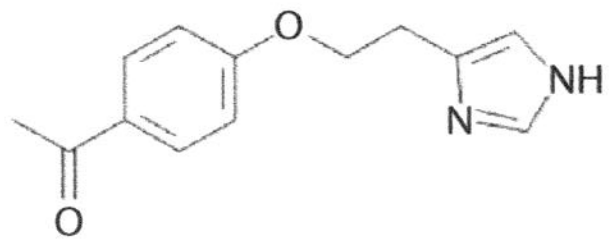	7.38	Stark et al. (2001)
42	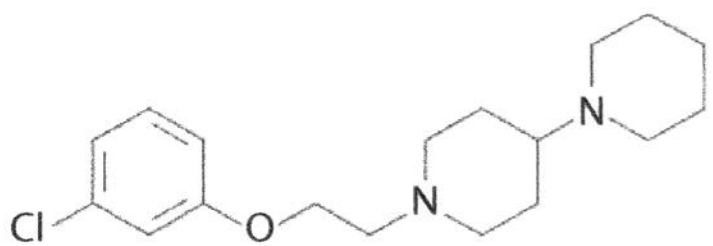	6.63	Lazewska et al. (2008)	95	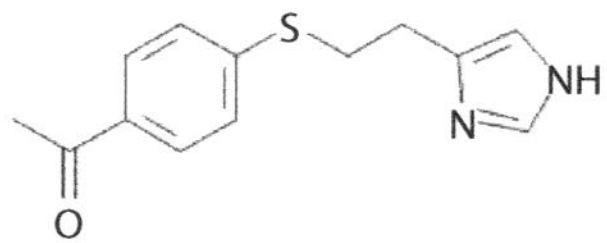	8.96	Stark et al. (2001)
43	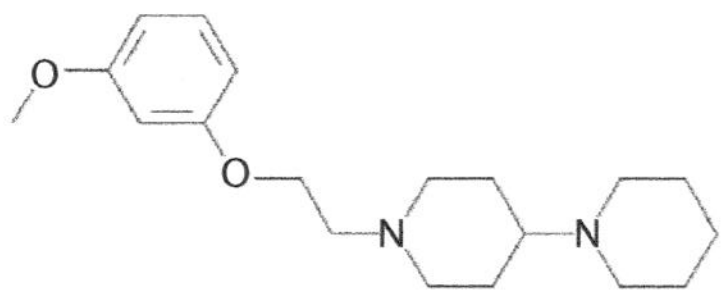	6.88	Lazewska et al. (2008)	96	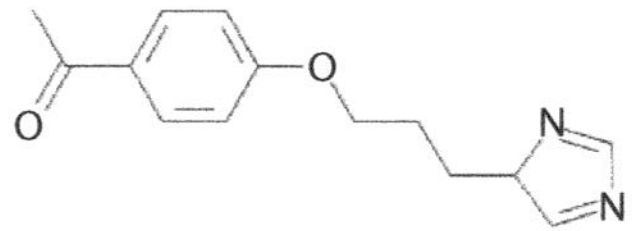	7.06	Stark et al. (2001)

(*Continued*)

TABLE 7.1 (*Continued*)
List of Selected Example Compounds for QSAR-Based Analyses

Compound Number	Structure	Biological Activity	References	Compound Number	Structure	Biological Activity	References
44	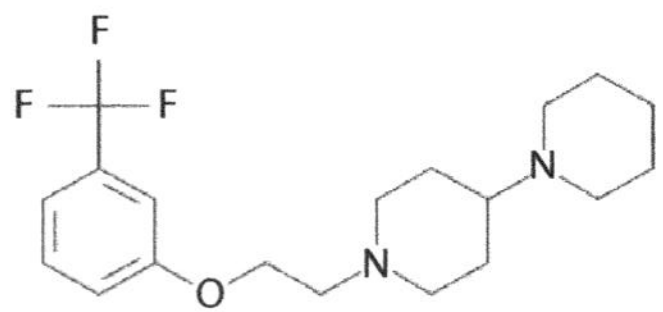	6.79	Lazewska et al. (2008)	97	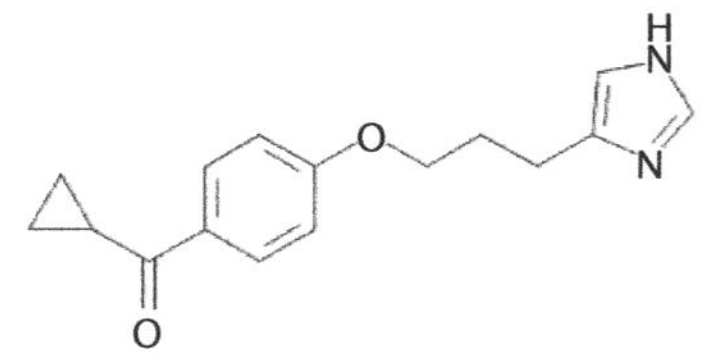	7.34	Stark et al. (2001)
45	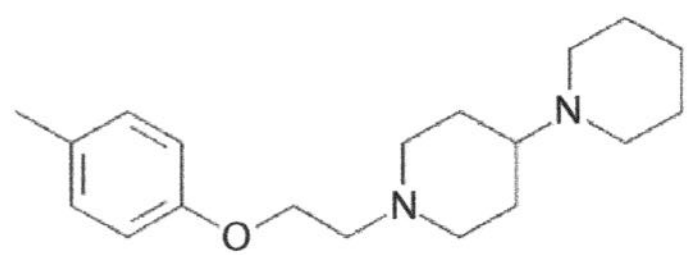	6.52	Lazewska et al. (2008)	98	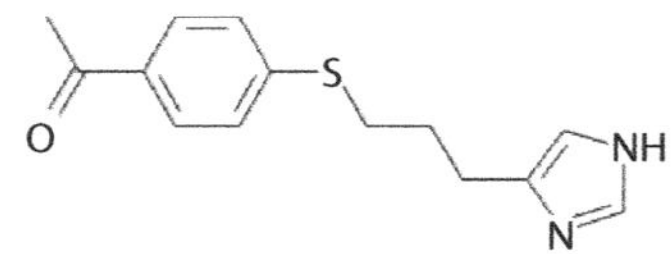	7.74	Stark et al. (2001)
46	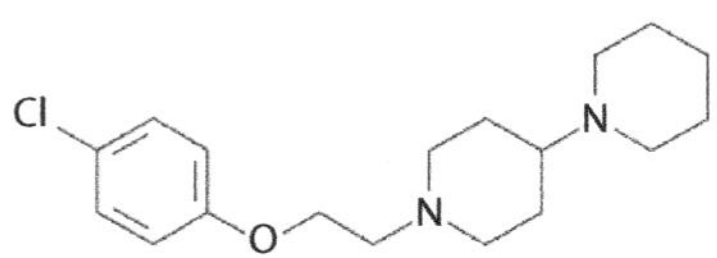	6.75	Lazewska et al. (2008)	99	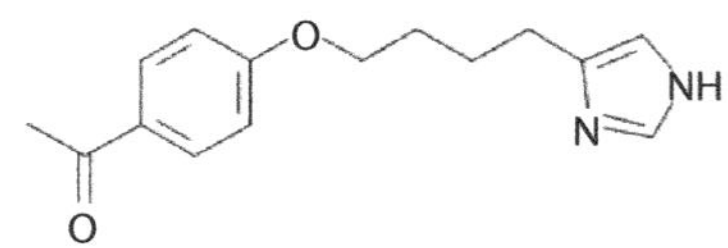	7.72	Stark et al. (2001)

(*Continued*)

TABLE 7.1 (*Continued*)
List of Selected Example Compounds for QSAR-Based Analyses

Compound Number	Structure	Biological Activity	References	Compound Number	Structure	Biological Activity	References
47	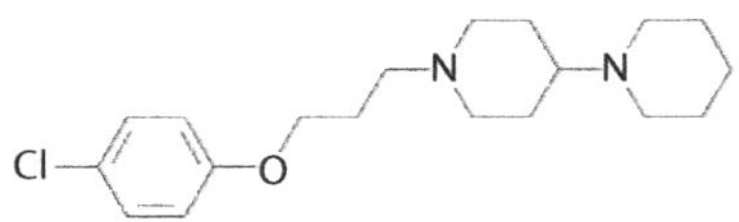	6.78	Lazewska et al. (2008)	100	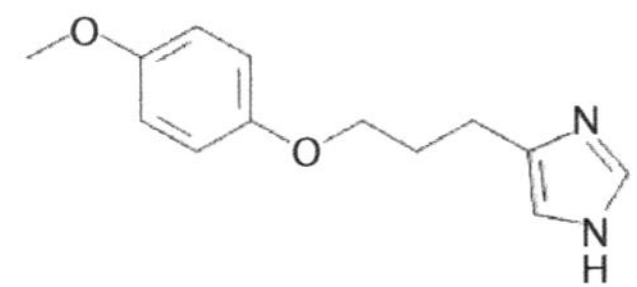	7.33	Stark et al. (2001)
48	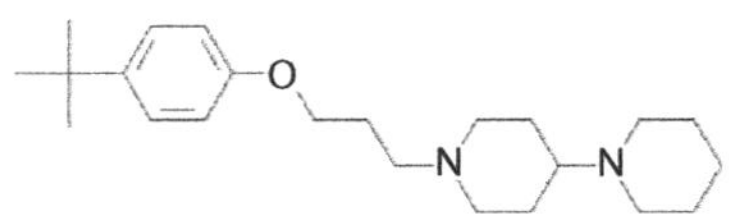	7	Lazewska et al. (2008)	101	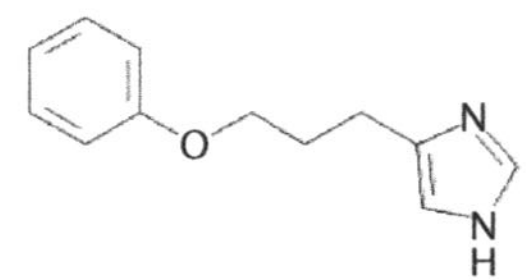	7.57	Stark et al. (2001)
49	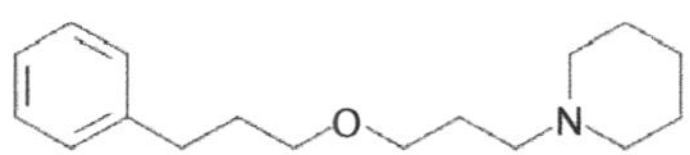	8.51	Lazewska et al. (2008)	102	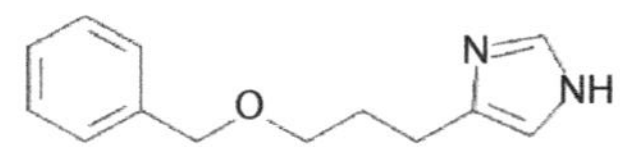	8.57	Stark et al. (2001)
50	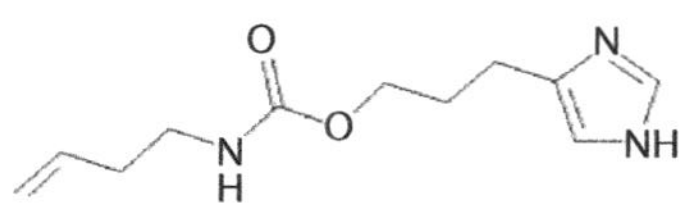	7.26	Wiecek et al. (2011)	103	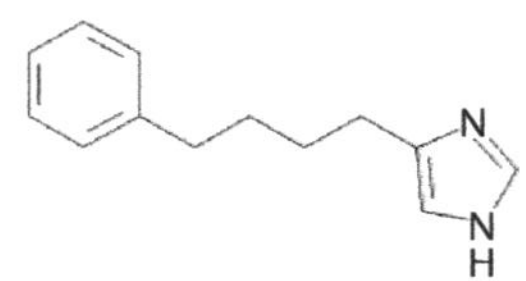	8.68	Stark et al. (2001)

(*Continued*)

TABLE 7.1 (*Continued*)
List of Selected Example Compounds for QSAR-Based Analyses

Compound Number	Structure	Biological Activity	References	Compound Number	Structure	Biological Activity	References
51	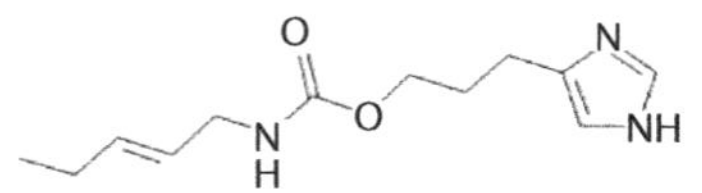	7.57	Wiecek et al. (2011)	104	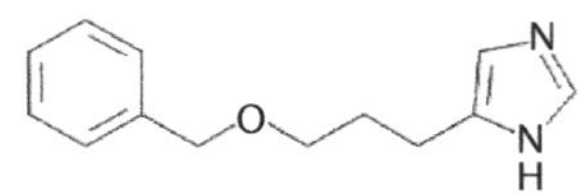	8.57	Stark et al. (2001)
52	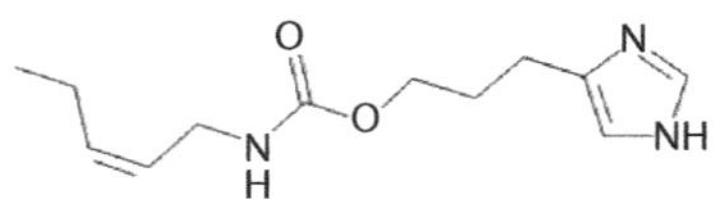	7.57	Wiecek et al. (2011)	105	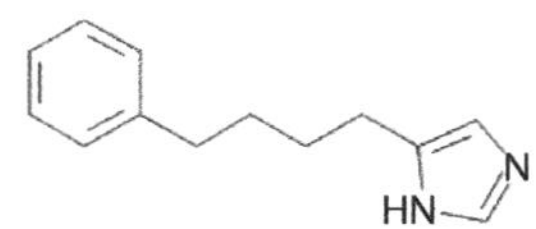	8.68	Stark et al. (2001)
53	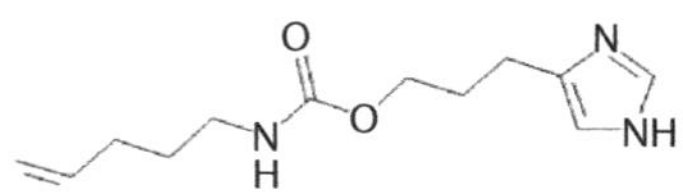	7.85	Wiecek et al. (2011)				

criterion for excluding the data point from the study. For instance, we were not sure how the activity for one of the compounds was obtained in the report by Amon et al. (2006), and hence, it was omitted from the analysis (Amon et al. 2006, Cowart et al. 2006).

By taking into account the above-mentioned measures, we are now confident to some extent that we have prepared an error-free data set ready to be proceeded into the next step.

The next step is to calculate the molecular descriptors for the compounds included in the data set. Energy minimization step was performed in HyperChem (version 8.0.8) that employs several methods including molecular mechanics, semiempirical, and *ab initio* among which AM1 semiempirical methodology was used in this practical example. The reason for using AM1 level of theory was its higher speed relative to more extensive *ab initio* calculations with comparable accuracy. Once the energy minimized conformations of the compounds were generated, they were used to calculate descriptors by Dragon (version 5.0), HyperChem (version 8.0.8), and ACDLabs suite of programs (version 2015.2.5). The number of calculated descriptors was almost close to 1,500 and to have a glimpse of what the values for the calculated descriptors look like, few of them are illustrated in Table 7.2 in an abbreviated fashion. Examples include constitutional, thermodynamic, topological, geometrical, and electronic descriptors.

Handling the large amount of descriptors for generating and evaluating all possible QSAR models is a computationally intensive task. Therefore, QSAR practitioners are recommended to take advantage of various algorithms to discard nonrelevant

TABLE 7.2
Representative Examples of Five Different Types of Descriptors

Descriptor Type		Constitutional Descriptor (MW)	Thermodynamic Descriptor (Log *P*)	Topological Descriptor (ZM1)	Geometrical Descriptor (Mor01m)	Electronic Descriptor (HOMO)
Descriptor Dimensionality		**0D**	**1D**	**2D**	**3D**	**4D**
Compound No.	1	171.32	1.11	50	95.1	−8.964
	2	185.35	1.58	54	111.94	−8.971
	3	199.38	1.97	58	130.144	−8.959
	4	213.41	2.38	64	149.713	−8.960
	5	227.44	2.77	68	170.646	−8.948
	⋮	⋮	⋮	⋮	⋮	⋮
	105	287.44	1.98	106	274.86	−8.982

Note:
MW is the molecular weight.
ZM1 is the first Zagreb index.
Mor01m is the signal 01/weighted by mass.
HOMO is the highest occupied molecular orbital.

and intercorrelated descriptors. For instance, in this table, HOMO parameter as an example of 4D descriptor should be discarded due to near constant values for the studied compounds. Such pretreatment of molecular descriptors results in a reduced and manageable set of descriptors.

Following the calculation of molecular parameters for the selected H3 antagonists, the pretreatment, train and test sets division, and descriptor selection procedures were performed. The data set was used in two different approaches. In the first approach, the entire data set was considered for the analysis, whereas in the second one, the data set was divided into imidazole- and nonimidazole-based compounds. Pretreatment of data sets was performed in three ways. In the first method, the molecular descriptors were processed on the basis of their variance and correlation coefficient cutoff values of 0.0001 and 0.99, respectively. In the second method, the parameters were normalized by satisfying the conditions of standard deviation of unity and mean value of zero. In the last approach, combination of above-mentioned two methods was applied for data pretreatment. Following the data pretreatment, the entire data set was divided into train (75%) and test (25%) sets using three algorithms, namely, Kennard-Stone, Euclidean distance, and activity-property. Then, the train set was subjected to descriptor selection procedure through GA–PLS. This process was repeated for 10 times to select the top-ranking parameters. The 50 top-ranking parameters for each run were selected, and a combined pool of descriptors were ranked on the basis of their frequencies of appearances. Figure 7.1 shows an example of 25 top-ranking parameters for imidazole- and nonimidazole-based compounds obtained from 10 GA-PLS runs.

Multiple linear regression (MLR) was used for finding the optimum relationship between top-ranking descriptors obtained from GA–PLS method and biological activity expressed by pK_i.

Because of the poor performances of the models developed using all 105 structures (combined imidazoles and nonimidazoles), they were not used for further analyses. Therefore, the data points were divided structurally into imidazole and nonimidazole derivatives. The MLR models were systematically generated and carefully inspected for finding the reliable QSAR models having one, two, three, and four variables (molecular descriptors). The inclusion of variables in the model was decided on the basis of the p values obtained for the coefficient values of the variables in the QSAR models. Presence of a variable was considered significant (different from zero) when the p value was less than 0.05.

As the validation is the most important step in developing predictive QSAR models, all the criteria for assessment of the generated models are going to be discussed in more details.

All generated MLR-based models were initially assessed for model selection based on R^2 (i.e., correlation coefficient) values using train and test sets. According to generally accepted threshold value of 0.6 for test set R^2, all models with R^2 of less

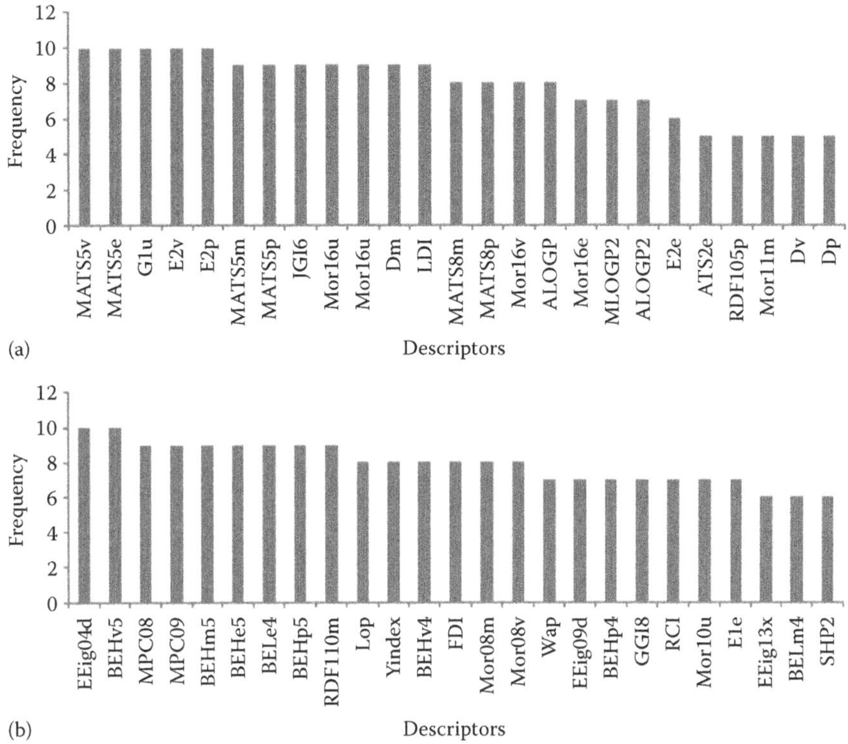

FIGURE 7.1 The GA–PLS selected parameters for imidazole (a) and nonimidazole (b) H3 receptor antagonists. Only 25 top-ranking molecular descriptors along with their frequencies are illustrated.

than 0.6 were not considered for further analyses. In this example, for H3 receptor antagonists, the analysis using all compounds did not satisfy the initial requirement of model validation. Therefore, the analyses were continued separately on imidazole- and nonimidazole-based compounds.

After selection of different QSAR models passing the initial assessment criteria (based on R^2), other validations were used for evaluating the reliability and predictivity of the generated QSAR models. Internal validation parameters such as leave-one-out Q^2, SDEP, r_m^2, and ${}^cR_p^2$ were calculated for the obtained models. Moreover, the Golbraikh and Tropsha criteria along with other external validation parameters such as Q^2_{F1}, Q^2_{F2}, Q^2_{F3}, CCC, and MAE were also calculated. The applicability domain procedure on the data set was performed as well for finding the possible outliers using Roy's MLR plus validation tool (Ambure et al. 2015).

The final selected QSAR models for imidazole and nonimidazole compounds along with their validation parameters are summarized below. The train and test

TABLE 7.3
Definitions of Molecular Descriptors for Equation 7.1

Parameter	Description	Class
MATS5v	Moran autocorrelation of lag 5 weighted by van der Waals volume	2D-autocorrelation
Mor16p	Signal 16/weighted by polarizability	3D-MoRSE descriptors
ZM2V	Second Zagreb by valence vertex degrees	Topological indices
Mor25m	Signal 25/weighted by mass	3D-MoRSE descriptors

sets division methodologies were based on Euclidean distance and activity–property based algorithms for imidazole and nonimidazole compounds, respectively:

QSAR model for imidazole-based compounds:

$$pK_i = 7.69952(\pm 0.05221) - 0.23539(\pm 0.06558)\,\mathrm{MATS\,5\,v} + 0.29445(\pm 0.07433)$$
$$\mathrm{Mor\,16\,p} - 0.28368(\pm 0.05638)\mathrm{ZM2V} + 0.2465(\pm 0.08543)\,\mathrm{Mor\,25\,m} \quad (7.1)$$

Table 7.3 shows the selected molecular descriptors for QSAR model generated for imidazole-based compounds. MATS5v belongs to 2D-autocorrelation descriptors. The 2D-autocorrelation descriptors are obtained on the basis of autocorrelation functions considering structural lags and lengths of the substructural fragments (Todeschini and Consonni 2008b). In MATS5v, the atomic van der Waals volumes are used as weighting coefficients of Moran Autocorrelation. This descriptor has a negative effect on the response values (i.e., pK_i).

3D-molecule representation of structures based on electron diffraction (3D-MoRSE) descriptors are calculated using information obtained from 3D atomic coordinates by the transform used in electron diffraction studies (Todeschini and Consonni 2008b) and weighted by the different physicochemical properties. In the current QSAR model, 3D-MoRSE descriptors are weighted by polarizability (Mor16p) and mass (Mor25m) of molecules and show positive impact on the endpoint values.

ZM2V is one of the topological descriptors originated from the 2D-graph representation of compounds in which the atoms and bonds are denoted by vertices and edges, respectively. This descriptor is obtained from the valence connectivity of the atoms using valence vertex degree. It is inversely related to the biological activity of the imidazole-based compounds.

QSAR model for nonimidazole-based compounds:

$$pK_i = 6.88172(\pm 0.05381) - 0.3405(\pm 0.10581)\,\mathrm{BEHm\,5} - 0.30056(\pm 0.09403)$$
$$\mathrm{BEFe\,5} + 0.7095(\pm 0.08032)\,\mathrm{BEHp\,4} + 0.37588(\pm 0.06164)Q_{\mathrm{mean}} \quad (7.2)$$

TABLE 7.4
Definitions of Molecular Descriptors for Equation 7.2

Parameter	Description	Class
BEHm5	Highest eigenvalue no. 5 of Burden matrix/weighted by atomic Sanderson masses	Eigenvalue-based descriptors
BEHe5	Highest eigenvalue no. 5 of Burden matrix/weighted by atomic Sanderson electronegativities	Eigenvalue-based descriptors
BEHp4	Highest eigenvalue no. 4 of Burden matrix/weighted by atomic Sanderson polarizability	Eigenvalue-based descriptors
Q_{mean}	Mean absolute charge (charge polarization)	Charge descriptors

Table 7.4 lists the molecular parameters selected for QSAR model using nonimidazole-based compounds. In the QSAR model generated for nonimidazole-based compounds, three similar molecular descriptors belonging to BCUT descriptors (so-called eigenvalue-based descriptors) were selected. These kinds of parameters are eigenvalues derived from connectivity information and atomic properties to form square symmetric matrix represented as a molecular graph. BCUT descriptors are weighted by different physicochemical properties such as mass, electronegativities, and polarizability (Gao 2001). BEHm5, BEHe5, and BEHp4 are the selected descriptors in the second QSAR model (Equation 7.2). As it can be seen in Equation 7.2, based on the model coefficients, BEHm5 and BEHe5 are inversely correlated with biological activity, whereas BEHp4 has a direct correlation with the response value.

Q_{mean} is the other descriptor in QSAR model (Equation 7.2) that refers to mean absolute charge. The positive effect of this descriptor implies that the greater value of the mean absolute charge of a given molecule leads to the increased biological activity. This is in agreement with BEHp4 parameter that is weighted by polarizability.

All the statistical criteria related to QSAR models (i.e., Equations 7.1 and 7.2) are listed in Table 7.5. The R^2 value of 0.8 is usually the minimum acceptable threshold for a QSAR involving biological data. However, it should be kept in mind that R^2 is not a particularly useful sole statistics as an indicator of goodness of fit. Therefore, considering errors of the predictions using criteria such as RMSEP is more reliable. Figure 7.2 demonstrates experimental versus predicted values of the biological activities of H3 imidazole- and nonimidazole-based antagonists.

To evaluate the collinearity of the selected parameters, the correlation matrix was generated for the selected descriptors and shown in Table 7.6. The table shows that there is no collinearity between the parameters greater than the threshold value of 0.8.

TABLE 7.5
The Summarized Statistics for QSAR Models along with Their Threshold Values Using Imidazole and Nonimidazole Based Compounds

	R^2	R^2_{adj}	Q^2_{LOO}	R^2_{test}	$\overline{r^2_{m}}_{LOO}$	Δr^2_{mLOO}	$\overline{r^2_{m}}_{test}$	Δr^2_{mtest}	$^{c}R^2_p$	RMSEP	Q^2_{F1}	Q^2_{F2}	Q^2_{F3}	CCC	MAE	K	K′	$\lvert R^2_0 - R'^2_0 \rvert$
Equation 7.1	0.68	0.64	0.57	0.65	0.46	0.14	0.53	0.23	0.63	0.35	0.64	0.63	0.53	0.79	0.19	0.99	1.01	0.13
Equation 7.2	0.7	0.67	0.58	0.69	0.47	0.15	0.53	0.2	0.66	0.39	0.69	0.68	0.57	0.81	0.27	0.99	1.01	0.18
Threshold			>0.5	>0.6	>0.5	<0.2	>0.5	<0.2						0.85		$0.9 \le k \le 1.1$	$0.9 \le k \le 1.1$	<0.3

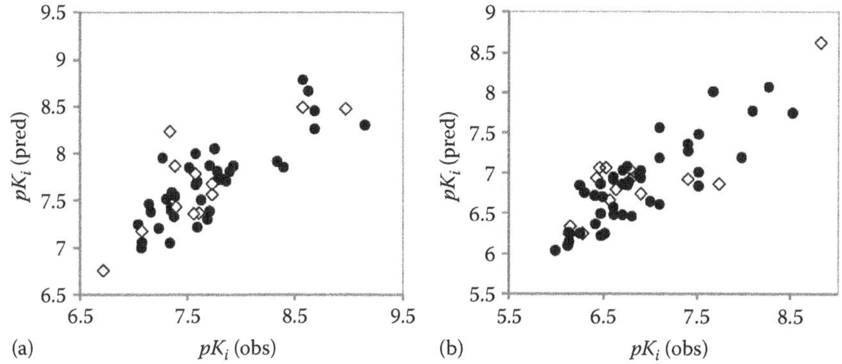

FIGURE 7.2 Plot of experimental versus predicted values of biological activities expressed as pK_i for imidazole- (a) and nonimidazole-based (b) H3 antagonists. Train set compounds are depicted as filled circles, whereas test sets are illustrated as open diamonds.

TABLE 7.6
Correlation Matrix for Selected Descriptors of both (A) Imidazole and (B) Nonimidazole-Based Compounds

	A					B			
	MATS5v	**Mor16p**	**ZM2V**	**Mor25m**		**BEHm5**	**BEHe5**	**BEHp4**	**Q_{mean}**
MATS5v	1				BEHm5	1			
Mor16p	0.479894	1			BEHe5	0.764291	1		
ZM2V	0.395129	0.298317	1		BEHp4	0.709407	0.634071	1	
Mor25m	0.607087	0.712346	0.360636	1	Q_{mean}	0.357884	0.333493	0.077536	1

The applicability domain analyses detected no outlier for imidazole-based compounds, whereas in the case of nonimidazoles one outlier (compound 80) was observed. However, a close inspection revealed that it was not a real outlier as ignoring this compound did not result in any improvement of the model in terms of statistical parameters and predictivity. Moreover, the other reason for not excluding this compound from the train set was avoiding the use of too narrow range of endpoint values.

8 Concluding Remarks

General concern on human health is one of the important challenges for the healthcare community. Despite the availability of large amount of pharmaceutical agents in the market, the need for designing new therapeutics is always in place because of the prevalence of drug resistance, undesirable side effects, as well as inefficiency of current drugs. Costly procedure of drug design and discovery is a driving force for taking advantage of computational approaches in this pipeline. *In silico* methods can accelerate the drug design processes by minimizing experimental approaches leading to less synthetic works, *in vitro*, and *in vivo* animal and human biomedical researches. In this regard, cheminformatics tools play an outstanding role in extracting and analyzing the data required for rational drug design. QSAR method as an inevitable component of medicinal chemistry finds the best mathematical model explaining a biological response in a given space of structural features. In other words, in this technique, the optimization of a particular feature is performed with the aim of indirect molecular design. The ultimate goal of a QSAR model is to provide beneficial information for rational lead optimization and to understand the mechanism of drug–target interactions, if possible. In addition, a reliable QSAR model can act as a virtual screening tool for finding the molecules with the desired biological activity among the chemical libraries. As mentioned in previous chapters, three main elements of a QSAR-based model are (1) experimental biological endpoints, (2) structural descriptors, and (3) statistical analyses. Each element can affect the quality of the QSAR model in several ways. The requirements for reporting a developed QSAR model should comply with the OECD guideline. All the assumptions as well as limitations of the developed QSAR model should be documented for allowing the reproducibility of the generated model and hence, assisting the decision-making power of medicinal chemists and other drug development team members. Otherwise, it may have a detrimental effect on the predictivity power of the final model rather than to be useful. Applicability domain as a model's confidence level of prediction should be determined in any QSAR study. Interpretability of a QSAR model is another issue that needs much attention. Use of a great number of molecular descriptors or noninterpretable descriptors leads to unreliable models in terms of certainty in the prediction and clarity in the mechanism of actions. There are countless publications in which the developed QSAR models have limited reliability and usefulness because of not following the OECD guideline. As a general rule, required amount of effort should be spent on improving the quality of QSAR models as much as possible using more evaluation tests such as "double cross-validation" just to give an example (Roy and Ambure 2016). At last but not least, one should bear in mind that a QSAR model is never free from errors and it is highly recommended to link QSAR analyses with experimental studies.

APPENDIX A

OECD Principles for the Validation, for Regulatory Purposes, of (Quantitative) Structure–Activity Relationship Models

These principles were agreed by OECD member countries at the 37th Joint Meeting of the Chemicals Committee and Working Party on Chemicals, Pesticides and Biotechnology in November 2004. The principles are intended to be read in conjunction with the associated explanatory notes, which were also agreed at the 37th Joint Meeting.

To facilitate the consideration of a (Q)SAR model for regulatory purposes, it should be associated with the following information:

1. A defined endpoint[1]
2. An unambiguous algorithm[2]
3. A defined domain of applicability[3]
4. Appropriate measures of goodness-of-fit, robustness, and predictivity[4]
5. A mechanistic interpretation, if possible[5]

NOTES

1. The intent of Principle 1 (defined endpoint) is to ensure clarity in the endpoint being predicted by a given model, since a given endpoint could be determined by different experimental protocols and under different experimental conditions. It is therefore important to identify the experimental system that is being modeled by the (Q)SAR. Further guidance is being developed regarding the interpretation of "defined endpoint". For example, a no-observed-effect level might be considered to be a defined endpoint in the sense that it is a defined information requirement of a given regulatory guideline, but cannot be regarded as a defined endpoint in the scientific sense of referring to a specific effect within a specific tissue/organ under specified conditions.
2. The intent of Principle 2 (unambiguous algorithm) is to ensure transparency in the model algorithm that generates predictions of an endpoint from information on chemical structure and/or physicochemical properties. It is recognized that, in the case of commercially developed models, this information is not always made publicly available. However, without this information, the performance of a model cannot be independently established, which is likely to represent a barrier for regulatory acceptance. The issue of reproducibility of the predictions is covered by this Principle, and will be explained further in the guidance material.
3. The need to define an applicability domain (Principle 3) expresses the fact that (Q)SARs are reductionist models, which are inevitably associated with limitations in terms of the types of chemical structures, physicochemical properties, and mechanisms of action for which the models can generate reliable predictions. Further work is recommended to define what types

of information are needed to define (Q)SAR applicability domains, and to develop appropriate methods for obtaining this information.

4. The revised Principle 4 (appropriate measures of goodness-of-fit, robustness, and predictivity) includes the intent of the original Setubal Principles 5 and 6. The wording of the principle is intended to simplify the overall set of principles, but not to lose the distinction between the internal performance of a model (as represented by goodness-of-fit and robustness) and the predictivity of a model (as determined by external validation). It is recommended that detailed guidance be developed on the approaches that could be used to provide appropriate measures of internal performance and predictivity. Further work is recommended to determine what constitutes external validation of (Q)SAR models.
5. It is recognized that it is not always possible, from a scientific viewpoint, to provide a mechanistic interpretation of a given (Q)SAR (Principle 5), or that there even be multiple mechanistic interpretations of a given model. The absence of a mechanistic interpretation for a model does not mean that a model is not potentially useful in the regulatory context. The intent of Principle 5 is not to reject models that have no apparent mechanistic basis, but to ensure that some consideration is given to the possibility of a mechanistic association between the descriptors used in a model and the endpoint being predicted, and to ensure that this association is documented.

References

Akutsu, T., and H. Nagamochi. 2013. Comparison and enumeration of chemical graphs. *Computational and Structural Biotechnology Journal*, 5:e201302004. doi:10.5936/csbj.201302004.

Alexander, G., and T. Alexander. 2002. Beware of q2! *Journal of Molecular Graphics and Modelling*, 20 (4):269–276. doi:10.1016/S1093-3263(01)00123-1.

Alkorta, I., F. Blanco, and J. Elguero. 2008. Application of Free-Wilson matrices to the analysis of the tautomerism and aromaticity of azapentalenes: A DFT study. *Tetrahedron*, 64 (17):3826–3836. doi:10.1016/j.tet.2008.01.141.

Ambure, P., R.B. Aher, A. Gajewicz, T. Puzyn, and K. Roy. 2015. "NanoBRIDGES" software: Open access tools to perform QSAR and nano-QSAR modeling. *Chemometrics and Intelligent Laboratory Systems*, 147:1–13. doi:10.1016/j.chemolab.2015.07.007.

Amon, M., X. Ligneau, J.C. Schwartz, and H. Stark. 2006. Fluorescent non-imidazole histamine H3 receptor ligands with nanomolar affinities. *Bioorganic Medicinal Chemistry Letters*, 16 (7):1938–1940. doi:10.1016/j.bmcl.2005.12.084.

Balaban, A.T. 1982. Highly discriminating distance-based topological index. *Chemical Physics Letters*, 89 (5):399–404.

Basant, N., C. Durante, M. Cocchi, and M.C. Menziani. 2012. Modeling the binding affinity of p38alpha MAP kinase inhibitors by partial least squares regression. *Chemical Biology & Drug Design*, 80 (3):455–470. doi:10.1111/j.1747-0285.2012.01419.x.

Baskin, I., and A. Varnek. 2008a. Building a chemical space based on fragment descriptors. *Combinational Chemistry & High Throughput Screening*, 11 (8):661–668.

Baskin, I., and A. Varnek. 2008b. Fragment descriptors in SAR/QSAR/QSPR studies, molecular similarity analysis and in virtual screening. In *Chemoinformatics Approaches to Virtual Screening*, A. Varnek and A. Tropsha (Eds.), pp. 1–43. Cambridge, UK: The Royal Society of Chemistry.

Becker, N., W. Werft, G. Toedt, P. Lichter, and A. Benner. 2009. penalizedSVM: A R-package for feature selection SVM classification. *Bioinformatics*, 25 (13):1711–1712.

Berrendero, J.R., A. Cuevas, and J.L. Torrecilla. 2016. The mRMR variable selection method: A comparative study for functional data. *Journal of Statistical Computation and Simulation*, 86 (5):891–907.

Breiman, L. 2001. Random forests. *Machine Learning*, 45 (1):5–32. doi:10.1023/A:1010933404324.

Brown, G., A. Pocock, M. Lujan, and M.-J. Zhao. 2012. Conditional likelihood maximisation: A unifying framework for information theoretic feature selection. *Journal of Machine Learning Research*, 13:27–66.

Calcagno, V., and C. de Mazancourt. 2010. Glmulti: An R package for easy automated model selection with (generalized) linear models. *Journal of Statistical Software*, 34 (12):1–29.

Canu, S., Y. Grandvalet, V. Guigue, and A. Rakotomamonjy. 2005. SVM and kernel methods Matlab toolbox. http://asi.insa-rouen.fr/~arakotom/toolbox/index.

Chen, L., H. Liu, J.P.A. Kocher, H. Li, and J. Chen. 2015. Glmgraph: An R package for variable selection and predictive modeling of structured genomic data. *Bioinformatics*, 31 (24):3991–3993.

Cherkasov, A., E.N. Muratov, D. Fourches, A. Varnek, I.I. Baskin, M. Cronin, J. Dearden et al. 2014. QSAR modeling: Where have you been? Where are you going to? *Journal of Medicinal Chemistry*, 57 (12):4977–5010. doi:10.1021/jm4004285.

Chirico, N., and P. Gramatica. 2011. Real external predictivity of QSAR models: How to evaluate it? Comparison of different validation criteria and proposal of using the concordance correlation coefficient. *Journal of Chemical Information and Modeling*, 51 (9):2320–2335. doi:10.1021/ci200211n.

Chirico, N., and P. Gramatica. 2012. Real external predictivity of QSAR models. Part 2. New intercomparable thresholds for different validation criteria and the need for scatter plot inspection. *Journal of Chemical Information and Modeling*, 52 (8):2044–2058. doi:10.1021/ci300084j.

Clementi, S., and S. Wold. 1995. How to choose the proper statistical method. In *Chemometric Methods in Molecular Design*, H. Waterbeemed (Ed.). New York: VCH.

Consonni, V., D. Ballabio, and R. Todeschini. 2009. Comments on the definition of the Q2 parameter for QSAR validation. *Journal of Chemical Information and Modeling*, 49 (7):1669–1678. doi:10.1021/ci900115y.

Consonni, V., D. Ballabio, and R. Todeschini. 2010. Evaluation of model predictive ability by external validation techniques. *Journal of Chemometrics*, 24 (3–4):194–201. doi:10.1002/cem.1290.

Consonni, V., R. Todeschini, and M. Pavan. 2002a. Structure/response correlations and similarity/diversity analysis by GETAWAY descriptors. 1. Theory of the novel 3D molecular descriptors. *Journal of Chemical Information and Computer Sciences*, 42 (3):682–692.

Consonni, V., R. Todeschini, M. Pavan, and P. Gramatica. 2002b. Structure/response correlations and similarity/diversity analysis by GETAWAY descriptors. 2. Application of the novel 3D molecular descriptors to QSAR/QSPR studies. *Journal of Chemical Information and Computer Sciences*, 42 (3):693–705.

Cowart, M., G.A. Gfesser, K. Bhatia, R. Esser, M. Sun, T.R. Miller, K. Krueger, D. Witte, T. A. Esbenshade, and A.A. Hancock. 2006. Fluorescent benzofuran histamine H(3) receptor antagonists with sub-nanomolar potency. *Inflammation Research*, 55 (Suppl 1): S47–S48. doi:10.1007/s00011-005-0036-y.

Craig, P.N. 1974. Comparison of the Hansch and Free-Wilson approaches to structure-activity correlation. In *Biological Correlations—The Hansch Approach*, pp. 115–129. Washington, DC: American Chemical Society.

Cronin, M.T.D., and T.W. Schultz. 2003. Pitfalls in QSAR. *Journal of Molecular Structure: THEOCHEM*, 622 (1–2):39–51. doi:10.1016/S0166-1280(02)00616-4.

Cruciani, G. 2005. *Molecular Interaction Fields-Applications in Drug Discovery and ADME Prediction*, R. Mannhold, H. Kubinyi and G. Folkers (Eds.), *Methods and Principles in Medicinal Chemistry*. Zurich, Switzerland: Wiley-VCH.

Dearden, J.C. 2016. The history and development of quantitative structure-Activity relationships (QSARs). *International Journal of Quantitative Structure-Property Relationships (IJQSPR)*, 1 (1):1–44. doi:10.4018/IJQSPR.2016010101.

Dearden, J.C. 2017. The use of topological indices in QSAR and QSPR modeling. In *Advances in QSAR Modeling: Applications in Pharmaceutical, Chemical, Food, Agricultural and Environmental Sciences*, K. Roy (Ed.), pp. 57–88. Cham, Switzerland: Springer International Publishing.

Dearden, J.C., M. Hewitt, and P.H. Rowe. 2015. QSAR study of some anti-hyperglycaemic sulphonylurea drugs. *SAR QSAR in Environmental Research*, 26 (6):439–448. doi:10.1080/1062936x.2015.1046189.

Dearden, J.C., M.T. Cronin, and K.L. Kaiser. 2009. How not to develop a quantitative structure-activity or structure-property relationship (QSAR/QSPR). *SAR QSAR in Environmental Research*, 20 (3–4):241–266. doi:10.1080/10629360902949567.

Deeb, O., and M. Goodarzi. 2010. Exploring QSARs for inhibitory activity of non-peptide HIV-1 protease inhibitors by GA-PLS and GA-SVM. *Chemical Biology & Drug Design*, 75 (5):506–514. doi:10.1111/j.1747-0285.2010.00953.x.

Dehmer, M.M., N.N. Barbarini, K.K. Varmuza, and A.A. Graber. 2010. Novel topological descriptors for analyzing biological networks. *BMC Structural Biology*, 10:18. doi:10.1186/1472-6807-10-18.

Demsar, J., T. Curk, A. Erjavec, C. Gorup, T. Hocevar, M. Milutinovic, M. Mozina et al. 2013. Orange: Data mining toolbox in python. *Journal of Machine Learning Research*, 14:2349–2353.

Deng, H., and G. Runger. 2013. Gene selection with guided regularized random forest. *Pattern Recognition*, 46 (12):3483–3489.

Devinyak, O., D. Havrylyuk, and R. Lesyk. 2014. 3D-MoRSE descriptors explained. *Journal of Molecular Graphic and Modelling*, 54:194–203. doi:10.1016/j.jmgm.2014.10.006.

Dey, T. 2013. modelSampler: An R tool for variable selection and model exploration in linear regression. *Journal of Data Science*, 11:343–370.

Dixon, S.L., J. Duan, E. Smith, C.D. Von Bargen, W. Sherman, and M.P. Repasky. 2016. AutoQSAR: An automated machine learning tool for best-practice quantitative structure-activity relationship modeling. *Future Medicinal Chemistry*, 8 (15):1825–1839. doi:10.4155/fmc-2016-0093.

Dong, J., D.S. Cao, H.Y. Miao, S. Liu, B.C. Deng, Y.H. Yun, N.N. Wang, A.P. Lu, W.B. Zeng, and A.F. Chen. 2015. ChemDes: An integrated web-based platform for molecular descriptor and fingerprint computation. *Journal of Cheminformatics*, 7:60. doi:10.1186/s13321-015-0109-z.

Dudek, A.Z., T. Arodz, and J. Galvez. 2006. Computational methods in developing quantitative structure-activity relationships (QSAR): A review. *Combinational Chemistry & High Throughput Screening*, 9 (3):213–228.

Dunn, W.J., J.H. Block, and R.S. Pearlman. 1986. *Partition Coefficient: Determination and Estimation*. New York: Pergamon Press.

Engel, T. 2012. Cheminformatics in diverse dimensions. In *Drug Design Strategies Computational Techniques and Applications*, T. Clark and L. Banting (Eds.), pp. 164–183. Cambridge, UK: The Royal Society of Chemistry.

Everitt, B., S. Landau, M. Leese, and D. Stahl. 2011. *Cluster Analysis*. 5th ed., *Wiley Series in Probability and Statistics*. Chichester, UK: Wiley.

Faulon, J.L., and A. Bender. 2010. *Handbook of Chemoinformatics Algorithms, Mathematical and Computational Biology Series*. Boca Raton, FL: Chapman & Hall/CRC.

Fröhlich, H., J.K. Wegner, and A. Zell. 2004. Towards optimal descriptor subset selection with support vector machines in classification and regression. *QSAR & Combinatorial Science*, 23 (5):311–318. doi:10.1002/qsar.200410011.

Fujita, T., and T. Ban. 1971. Structure-activity study of phenethylamines as substrates of biosynthetic enzymes of sympathetic transmitters. *Journal of Medicinal Chemistry* 14 (2):148–152.

Funatsu, K., T. Miyao, and M. Arakawa. 2011. Systematic generation of chemical structures for rational drug design based on QSAR models. *Current Computer-Aided Drug Design*, 7 (1):1–9.

Gao, H. 2001. Application of BCUT metrics and genetic algorithm in binary QSAR analysis. *Journal of Chemical Information and Computer Sciences*, 41 (2):402–407.

Garg, P., and J. Verma. 2006. In silico prediction of blood brain barrier permeability: An artificial neural network model. *Journal of Chemical Information and Modeling*, 46 (1):289–297. doi:10.1021/ci050303i.

Gasteiger, J., and J. Zupan. 1993. Neural networks in chemistry. *Angewandte Chemie-International Edition in English*, 32 (4):503–527. doi:10.1002/anie.199305031.

Gedeck, P., B. Rohde, and C. Bartels. 2006. QSAR–how good is it in practice? Comparison of descriptor sets on an unbiased cross section of corporate data sets. *Journal of Chemical Information and Modeling*, 46 (5):1924–1936. doi:10.1021/ci050413p.

Genuer, R., J.M. Poggi, and C. Tuleau-Malot. 2010. Variable selection using random forests. *Pattern Recognition Letters*, 31 (14):2225–2236. doi:10.1016/j.patrec.2010.03.014.

Genuer, R., J.M. Poggi, and C. Tuleau-Malo. 2015. VSURF: An R package for variable selection using random forests. *The R Journal*, 7 (2):19–33.

Ghandadi, M., A. Shayanfar, M. Hamzeh-Mivehroud, and A. Jouyban. 2014. Quantitative structure activity relationship and docking studies of imidazole-based derivatives as P-glycoprotein inhibitors. *Medicinal Chemistry Research*, 23 (11):4700–4712. doi:10.1007/s00044-014-1029-6.

Gobbi, A., and M.L. Lee. 2003. DISE: Directed sphere exclusion. *Journal of Chemical Information and Computer Sciences*, 43 (1):317–323. doi:10.1021/ci025554v.

Golbraikh, A., X.S. Wang, H. Zhu, and A. Tropsha. 2012. Predictive QSAR modeling: Methods and applications in drug discovery and chemical risk assessment. In *Handbook of Computational Chemistry*, J. Leszczynski (Ed.), pp. 1309–1342. Dordrecht, the Netherlands: Springer.

Goodarzi, M., Y. Vander Heyden, and S. Funar-Timofei. 2013. Towards better understanding of feature-selection or reduction techniques for quantitative structure–Activity relationship models. *TrAC Trends in Analytical Chemistry*, 42:49–63. doi:10.1016/j.trac.2012.09.008.

Gozalbes, R., and A. Pineda-Lucena. 2011. Small molecule databases and chemical descriptors useful in chemoinformatics: An overview. *Combinatorial Chemistry & High Throughput Screening*, 14 (6):548–458.

Gramatica, P. 2007. Principles of QSAR models validation: Internal and external. *QSAR & Combinatorial Science*, 26 (5):694–701. doi:10.1002/qsar.200610151.

Gramatica, P., and A. Sangion. 2016. A historical excursus on the statistical validation parameters for QSAR models: A clarification concerning metrics and terminology. *Journal of Chemical Information and Modeling*, 56 (6):1127–1131. doi:10.1021/acs.jcim.6b00088.

Gramatica, P., S. Cassani, P.P. Roy, S. Kovarich, C.W. Yap, and E. Papa. 2012. QSAR modeling is not "Push a button and find a correlation": A case study of toxicity of (Benzo-)triazoles on algae. *Molecular Informatics*, 31 (11–12):817–835. doi:10.1002/minf.201200075.

Gramatica, P., N. Chirico, E. Papa, S. Cassani, and S. Kovarich. 2013. QSARINS: A new software for the development, analysis, and validation of QSAR MLR models. *Journal of Computational Chemistry*, 34 (24):2121–2132. doi:10.1002/jcc.23361.

Graovac, A., I. Gutman, N. Trinajstić, and T. Živković. 1972. Graph theory and molecular orbitals. *Theoretical Chemistry Accounts: Theory, Computation, and Modeling*, 26 (1):67–78. doi:10.1007/bf00527654.

Grassmann, S., J. Apelt, W. Sippl, X. Ligneau, H.H. Pertz, Y.H. Zhao, J.M. Arrang et al. 2003. Imidazole derivatives as a novel class of hybrid compounds with inhibitory histamine N-methyltransferase potencies and histamine hH3 receptor affinities. *Bioorganic & Medicinal Chemistry*, 11 (10):2163–2174.

Guyon, I., and A. Elisseeff. 2003. An introduction to variable and feature selection. *Journal of Machine Learning Research*, 3:1157–1182.

Hall, L.H., and L.B. Kier. 2007. The molecular connectivity chi indexes and kappa shape indexes in structure-property modeling. In *Reviews in Computational Chemistry*, pp. 367–422. Hoboken, NJ: John Wiley & Sons.

Hall, M., E. Frank, G. Holmes, B. Pfahringer, P. Reutemann, and I.H. Witten. 2009. The WEKA data mining software: An update. *SIGKDD Explorations*, 11 (1):10–18.

Hammett, L.P. 1937. The effect of structure upon the reactions of organic compounds. Benzene derivatives. *Journal of the American Chemical Society*, 59 (1):96–103. doi:10.1021/ja01280a022.

Hamzeh-Mivehroud, M., B. Sokouti, and S. Dastmalchi. 2015. An introduction to the basic concepts in QSAR-aided drug design. In *Quantitative Structure-Activity Relationships in Drug Design, Predictive Toxicology, and Risk Assessment*, K. Roy (Ed.), pp. 1–47. Hershey, PA: IGI Global.

Hamzeh-Mivehroud, M., H. Moghaddas-Sani, M. Rahbar-Shahrouziasl, and S. Dastmalchi. 2015. Identifying key interactions stabilizing DOF zinc finger-DNA complexes using in silico approaches. *Journal of Theoretical Biology*, 382:150–159. doi:10.1016/j.jtbi.2015.06.013.

Hamzeh-Mivehroud, M., S. Rahmani, M.A. Feizi, S. Dastmalchi, and M.R. Rashidi. 2014. In vitro and in silico studies to explore structural features of flavonoids for aldehyde oxidase inhibition. *Archiv der Pharmazie (Weinheim)*, 347 (10):738–747. doi:10.1002/ardp.201400076.

Hansch, C., P.P. Maloney, T. Fujita, and R.M. Muir. 1962. Correlation of biological activity of phenoxyacetic acids with Hammett substituent constants and partition coefficients. *Nature*, 194 (4824):178–180.

Helguera, A.M., R.D. Combes, M.P. Gonzalez, and M.N. Cordeiro. 2008. Applications of 2D descriptors in drug design: A DRAGON tale. *Current Topics in Medical Chemistry*, 8 (18):1628–1655.

Hofmann, M., and R. Klinkenberg. 2013. *RapidMiner: Data Mining Use Cases and Business Analytics Applications*. Boca Raton, FL: CRC Press.

Hudson, B.D., R.M. Hyde, E. Rahr, J. Wood, and J. Osman. 1996. Parameter based methods for compound selection from chemical databases. *Quantitative Structure-Activity Relationships*, 15 (4):285–289. doi:10.1002/qsar.19960150402.

Jaworska, J., N. Nikolova-Jeliazkova, and T. Aldenberg. 2005. QSAR applicability domain estimation by projection of the training set descriptor space: A review. *Alternatives to Laboratory Animals*, 33 (5):445–459.

Jerome, F., H. Trevor, and T. Robert. 2010. Regularization paths for generalized linear models via coordinate descent. *Journal of Statistical Software*, 33 (1):1–22.

Karelson, M., V.S. Lobanov, and A.R. Katritzky. 1996. Quantum-chemical descriptors in QSAR/QSPR studies. *Chemical Reviews*, 96 (3):1027–1044.

Karelson, M., U. Maran, Y. Wang, and A.R. Katritzky. 1999. QSPR and QSAR models derived using large molecular descriptor spaces. A review of CODESSA applications. *Collection of Czechoslovak Chemical Communications*, 64:1551–1571. Institute of Organic Chemistry and Biochemistry.

Katritzky, A.R., O.V. Kulshyn, I. Stoyanova-Slavova, D.A. Dobchev, M. Kuanar, D.C. Fara, and M. Karelson. 2006. Antimalarial activity: A QSAR modeling using CODESSA PRO software. *Bioorganic & Medicinal Chemistry*, 14 (7):2333–2357. doi:10.1016/j.bmc.2005.11.015.

Katritzky, A.R., V.S. Lobanov, and M. Karelson. 1995. QSPR: The correlation and quantitative prediction of chemical and physical properties from structure. *Chemical Society Reviews* 24 (4):279–287. doi:10.1039/CS9952400279.

Kennard, R.W., and L.A. Stone. 1969. Computer aided design of experiments. *Technometrics* 11 (1):137–148. doi:10.2307/1266770.

Kennedy, J., and R.C. Eberhart. 1994. Particle swarm optimization. *Proceedings IEEE International Conference on Neural Networks*, 4:1942–1948.

Khan, A.U. 2016. Descriptors and their selection methods in QSAR analysis: Paradigm for drug design. *Drug Discovery Today*, 21:1291–1302. doi:10.1016/j.drudis.2016.06.013.

Kier, L.B., and L.H. Hall. 1986. *Molecular Connectivity in Structure-Activity Analysis*. New York: Wiley.

Kirchmair, J., A.H. Goller, D. Lang, J. Kunze, B. Testa, I.D. Wilson, R.C. Glen, and G. Schneider. 2015. Predicting drug metabolism: Experiment and/or computation? *Nature Reviews Drug Discovery*, 14 (6):387–404. doi:10.1038/nrd4581.

Knox, C., V. Law, T. Jewison, P. Liu, S. Ly, A. Frolkis, A. Pon et al. 2011. DrugBank 3.0: A comprehensive resource for "omics" research on drugs. *Nucleic Acids Research*, 39 (Database issue):D1035–D1041. doi:10.1093/nar/gkq1126.

Kubinyi, H. 1988. Free Wilson analysis. Theory, applications and its relationship to Hansch analysis. *Quantitative Structure-Activity Relationships*, 7 (3):121–133. doi:10.1002/qsar.19880070303.

Kubinyi, H. 1993. *QSAR: Hansch Analysis and Related Approaches, Methods and Principles in Medicinal Chemistry*. Weinheim, Germany: VCH.

Kuhn, M. 2013. Caret: Classification and regression training (R package version 5.17-7). http://CRAN.R-project.org/package=caret.

Kursa, M.B., and W.R. Rudnick. 2010. Feature selection with the Boruta package. *Journal of Statistical Software*, 36 (11):1–13.

Lagorce, D., C. Reynes, A.C. Camproux, M.A. Miteva, O. Sperandio, and B.O. Villoutreix. 2011. In silico ADME/tox predictions. In *ADMET for Medicinal Chemists*, pp. 29–124. Hoboken, NJ: John Wiley & Sons.

Law, V., C. Knox, Y. Djoumbou, T. Jewison, A.C. Guo, Y. Liu, A. Maciejewski et al. 2014. DrugBank 4.0: Shedding new light on drug metabolism. *Nucleic Acids Research*, 42 (Database issue):D1091–D1097. doi:10.1093/nar/gkt1068.

Lazewska, D., K. Kuder, X. Ligneau, J.C. Schwartz, W. Schunack, H. Stark, and K. Kiec-Kononowicz. 2008. Piperidine variations in search for non-imidazole histamine H(3) receptor ligands. *Bioorganic & Medicinal Chemistry*, 16 (18):8729–8736. doi:10.1016/j.bmc.2008.07.071.

Lazewska, D., X. Ligneau, J.C. Schwartz, W. Schunack, H. Stark, and K. Kiec-Kononowicz. 2006. Ether derivatives of 3-piperidinopropan-1-ol as non-imidazole histamine H3 receptor antagonists. *Bioorganic & Medicinal Chemistry*, 14 (10):3522–3529. doi:10.1016/j.bmc.2006.01.013.

Lazewska, D., M. Wiecek, X. Ligneau, T. Kottke, L. Weizel, R. Seifert, W. Schunack, H. Stark, and K. Kiec-Kononowicz. 2009. Histamine H3 and H4 receptor affinity of branched 3-(1H-imidazol-4-yl) propyl N-alkylcarbamates. *Bioorganic & Medicinal Chemistry Letters*, 19 (23):6682–6685. doi:10.1016/j.bmcl.2009.10.005.

Leach, A.R., and V.J. Gillet. 2007. *An Introduction To Chemoinformatics*. Springer.

Leardi, R. 2000. Application of genetic algorithm-PLS for feature selection in spectral data sets. *Journal of Chemometrics*, 14:643–655.

Lewis, R.A., and D. Wood. 2014. Modern 2D QSAR for drug discovery. *Wiley Interdisciplinary Reviews: Computational Molecular Science*, 4 (6):505–522. doi:10.1002/wcms.1187.

Li, H., Q. Xu, and Y. Liang. 2014. libPLS: An integrated library for partial least squares regression and discriminant analysis. http://www.libpls.net.

Liaw, A., and M. Wiener. 2002. Classification and regression by randomForest. *R News* 2 (3):18–22.

Ligneau, X., S. Morisset, J. Tardivel-Lacombe, F. Gbahou, C.R. Ganellin, H. Stark, W. Schunack, J.C. Schwartz, and J.M. Arrang. 2000. Distinct pharmacology of rat and human histamine H(3) receptors: Role of two amino acids in the third transmembrane domain. *British Journal of Pharmacology*, 131 (7):1247–1750. doi:10.1038/sj.bjp.0703712.

Lin, L.I. 1989. A concordance correlation coefficient to evaluate reproducibility. *Biometrics* 45 (1):255–268.

Liu, H., E. Papa, and P. Gramatica. 2006. QSAR prediction of estrogen activity for a large set of diverse chemicals under the guidance of OECD principles. *Chemical Research in Toxicology*, 19 (11):1540–1548. doi:10.1021/tx0601509.

Liu, S.S., H.L. Liu, C.S. Yin, and L.S. Wang. 2003. VSMP: A novel variable selection and modeling method based on the prediction. *Journal of Chemical Information and Computer Sciences*, 43 (3):964–969. doi:10.1021/ci020377j.

Liu, Y. 2004. A comparative study on feature selection methods for drug discovery. *Journal of Chemical Information and Computer Sciences*, 44 (5):1823–1828. doi:10.1021/ci049875d.

Maggiora, G.M. 2006. On outliers and activity cliffs–why QSAR often disappoints. *Journal of Chemical Information and Modeling*, 46 (4):1535. doi:10.1021/ci060117s.

Martin, T.M., P. Harten, D.M. Young, E.N. Muratov, A. Golbraikh, H. Zhu, and A. Tropsha. 2012. Does rational selection of training and test sets improve the outcome of QSAR modeling? *Journal of Chemical Information and Modeling*, 52 (10):2570–2578. doi:10.1021/ci300338w.

McLeod, A.I., and C. Xu. 2014. bestglm: Best subset GLM. https://CRAN.R-project.org/package=bestglm.

Mehmood, T., K.H. Liland, L. Snipen, and S. Saebo. 2012. A review of variable selection methods in partial least squares regression. *Chemometrics and Intelligent Laboratory Systems*, 118:62–69. doi:10.1016/j.chemolab.2012.07.010.

Meissner, M., M. Schmuker, and G. Schneider. 2006. Optimized particle swarm optimization (OPSO) and its application to artificial neural network training. *BMC Bioinformatics*, 7:125. doi:10.1186/1471-2105-7-125.

Miko, T., X. Ligneau, H.H. Pertz, J.M. Arrang, C.R. Ganellin, J.C. Schwartz, W. Schunack, and H. Stark. 2004. Structural variations of 1-(4-(phenoxymethyl)benzyl) piperidines as nonimidazole histamine H3 receptor antagonists. *Bioorganic & Medicinal Chemistry*, 12 (10):2727–2736. doi:10.1016/j.bmc.2004.03.009.

Mitchell, J.B.O. 2014. Machine learning methods in chemoinformatics. *Wiley Interdisciplinary Reviews: Computational Molecular Science*, 4 (5):468–481. doi:10.1002/wcms.1183.

Nandy, A., S. Kar, and K. Roy. 2013. Linear discriminant analysis for skin sensitisation potential of diverse organic chemicals. *Molecular Simulation*, 39 (6):432–441. doi:10.1080/08927022.2012.738421.

OECD. 2004. Report from the Expert Group on (Quantitative) Structure-Activity Relationships [(Q)SARs] on the Principles for the Validation of (Q)SARs, Series on Testing and Assessment, Paris.

Pliska, V. 2010. Thermodynamic descriptors, profiles and driving forces in membrane receptor-ligand interactions. *Journal of Receptors and Signal Transduction Research*, 30 (6):454–468. doi:10.3109/10799893.2010.515594.

Randic, M. 1975. On characterization of molecular branching. *Journal of the American Chemical Society*, 97 (23):6609–6615. doi:10.1021/ja00856a001.

Rocha, G.B., R.O. Freire, A.M. Simas, and J.J. Stewart. 2006. RM1: A reparameterization of AM1 for H, C, N, O, P, S, F, Cl, Br, and I. *Journal of Computational Chemistry*, 27 (10):1101–1111. doi:10.1002/jcc.20425.

Roy, K., and I. Mitra. 2012. Electrotopological state atom (E-state) index in drug design, QSAR, property prediction and toxicity assessment. *Current Computer Aided Drug Design*, 8 (2):135–158.

Roy, K., and P. Ambure. 2016. The "double cross-validation" software tool for MLR QSAR model development. *Chemometrics and Intelligent Laboratory Systems*, 159:108–126. doi:10.1016/j.chemolab.2016.10.009.

Roy, K., P. Chakraborty, I. Mitra, P.K. Ojha, S. Kar, and R.N. Das. 2013. Some case studies on application of "rm2" metrics for judging quality of quantitative structure–activity relationship predictions: Emphasis on scaling of response data. *Journal of Computational Chemistry*, 34 (12):1071–1082. doi:10.1002/jcc.23231.

Roy, K., S. Kar, and P. Ambure. 2015. On a simple approach for determining applicability domain of QSAR models. *Chemometrics and Intelligent Laboratory Systems*, 145:22–29. doi:10.1016/j.chemolab.2015.04.013.

Roy, K., S. Kar, and R.N. Das. 2015a. *A Primer on QSAR/QSPR Modeling Fundamental Concepts*. Cham, Switzerland: Springer.

Roy, K., S. Kar, and R.N. Das. 2015b. Chapter 7-Validation of QSAR models. In *Understanding the Basics of QSAR for Applications in Pharmaceutical Sciences and Risk Assessment*, pp. 231–289. Boston, MA: Academic Press.

Roy, P.P., J.T. Leonard, and K. Roy. 2008. Exploring the impact of size of training sets for the development of predictive QSAR models. *Chemometrics and Intelligent Laboratory Systems*, 90 (1):31–42. doi:10.1016/j.chemolab.2007.07.004.

Sagrado, S., and M.T. Cronin. 2008. Application of the modelling power approach to variable subset selection for GA-PLS QSAR models. *Analytica Chimica Acta*, 609 (2):169–174. doi:10.1016/j.aca.2008.01.013.

Saptoro, A., M.O. Tadé, and H. Vuthaluru. 2012. A modified Kennard-stone algorithm for optimal division of data for developing artificial neural network models. In *Chemical Product and Process Modeling.*

Schüürmann, G., R.-U. Ebert, J. Chen, B. Wang, and R. Kühne. 2008. External validation and prediction employing the predictive squared correlation coefficient—Test set activity mean vs training set activity mean. *Journal of Chemical Information and Modeling*, 48 (11):2140–2145. doi:10.1021/ci800253u.

Sciabola, S., R.V. Stanton, T.L. Johnson, and H. Xi. 2011. Application of Free-Wilson selectivity analysis for combinatorial library design. *Methods in Molecular Biology*, 685:91–109. doi:10.1007/978-1-60761-931-4_5.

Sciabola, S., R.V. Stanton, S. Wittkopp, S. Wildman, D. Moshinsky, S. Potluri, and H. Xi. 2008. Predicting kinase selectivity profiles using Free-Wilson QSAR analysis. *Journal of Chemical Information and Modeling*, 48 (9):1851–1867. doi:10.1021/ci800138n.

Scior, T., J.L. Medina-Franco, Q.T. Do, K. Martinez-Mayorga, J.A. Yunes Rojas, and P. Bernard. 2009. How to recognize and workaround pitfalls in QSAR studies: A critical review. *Current Medicinal Chemistry*, 16 (32):4297–4313.

Shahlaei, M. 2013. Descriptor selection methods in quantitative structure-activity relationship studies: A review study. *Chemical Reviews*, 113 (10):8093–8103. doi:10.1021/cr3004339.

Shamsipur, M., V. Zare-Shahabadi, B. Hemmateenejad, and M. Akhond. 2009. An efficient variable selection method based on the use of external memory in ant colony optimization. Application to QSAR/QSPR studies. *Analytica Chimica Acta*, 646 (1–2):39–46. doi:10.1016/j.aca.2009.05.005.

Shi, L.M., H. Fang, W. Tong, J. Wu, R. Perkins, R.M. Blair, W.S. Branham, S.L. Dial, C.L. Moland, and D.M. Sheehan. 2001. QSAR models using a large diverse set of estrogens. *Journal of Chemical Information and Computer Sciences*, 41 (1):186–195. doi:10.1021/ci000066d.

Singh, H., S. Singh, D. Singla, S.M. Agarwal, and G.P. Raghava. 2015. QSAR based model for discriminating EGFR inhibitors and non-inhibitors using random forest. *Biology Direct*, 10:10. doi:10.1186/s13062-015-0046-9.

Singla, D., S.K. Dhanda, J.S. Chauhan, A. Bhardwaj, S.K. Brahmachari, and G.P. Raghava. 2013. Open source software and web services for designing therapeutic molecules. *Current Topics in Medicinal Chemistry*, 13 (10):1172–1191.

Sokouti, B., F. Rezvan, and S. Dastmalchi. 2015. Applying random forest and subtractive fuzzy c-means clustering techniques for the development of a novel G protein-coupled receptor discrimination method using pseudo amino acid compositions. *Molecular BioSystems*, 11 (8):2364–2372. doi:10.1039/c5mb00192g.

Somol, P., P. Vácha, S. Mikeš, J. Hora, and P. Pudil. 2010. Introduction to feature selection Toolbox 3—The C++ library for subset search, data modeling and classification. In *Research Report for Institute of Information Theory and Automation*. Academy of Sciences of the Czech Republic.

Soto, A.J., R.L. Cecchini, G.E. Vazquez, and I. Ponzoni. 2009. Multi-objective feature selection in QSAR using a machine learning approach. *QSAR & Combinatorial Science*, 28 (11–12):1509–1523. doi:10.1002/qsar.200960053.

Stark, H., W. Sippl, X. Ligneau, J.M. Arrang, C.R. Ganellin, J.C. Schwartz, and W. Schunack. 2001. Different antagonist binding properties of human and rat histamine H3 receptors. *Bioorganic & Medicinal Chemistry Letters*, 11 (7):951–954.

Stewart, J.J. 2007. Optimization of parameters for semiempirical methods V: Modification of NDDO approximations and application to 70 elements. *Journal of Molecular Modeling*, 13 (12):1173–1213. doi:10.1007/s00894-007-0233-4.

Sukumar, N., G. Prabhu, and P. Saha. 2014. Applications of genetic algorithms in QSAR/QSPR modeling. In *Applications of Metaheuristics in Process Engineering*, J. Valadi and P. Siarry (Eds.), pp. 315–324. Cham, Switzerland: Springer International Publishing.

Svetnik, V., A. Liaw, C. Tong, J.C. Culberson, R.P. Sheridan, and B.P. Feuston. 2003. Random forest: A classification and regression tool for compound classification and QSAR modeling. *Journal of Chemical Information and Computer Sciences*, 43 (6):1947–1958. doi:10.1021/ci034160g.

Szantai-Kis, C., I. Kovesdi, G. Keri, and L. Orfi. 2003. Validation subset selections for extrapolation oriented QSPAR models. *Molecular Diversity*, 7 (1):37–43.

Team, R.C. 2015. R: A language and environment for statistical computing. https://www.R-project.org/.

Terada, Y., and K. Nanya. 2000. Free-Wilson analysis of the antibacterial activity of fluoronaphthyridines against various microbes. A new application of indicator variables. *Pharmazie*, 55 (2):133–135.

Tetko, I.V., V.V. Kovalishyn, and D.J. Livingstone. 2001. Volume learning algorithm artificial neural networks for 3D QSAR studies. *Journal of Medicinal Chemistry*, 44 (15):2411–2420. doi:10.1021/jm010858e.

TheMathWorksInc. 2016. MATLAB release 2016b. http://www.matlab.com.

Tichy, M., and M. Rucki. 2009. Validation of QSAR models for legislative purposes. *Interdisciplinary Toxicology*, 2 (3):184–186. doi:10.2478/v10102-009-0014-2.

Todeschini, R., and V. Consonni. 2008a. Descriptors from molecular geometry. In *Handbook of Chemoinformatics*, pp. 1004–1033. Weinheim, Germany: Wiley-VCH Verlag GmbH.

Todeschini, R., and V. Consonni. 2008b. *Handbook of Molecular Descriptors, Methods and Principles in Medicinal Chemistry*. Weinheim, Germany: Wiley-VCH Verlag GmbH.

Todeschini, R., and V. Consonni. 2009. *Molecular Descriptors for Chemoinformatics, Volumes I & II*. Weinheim, Germany: Wiley-VCH.

Todeschini, R., and V. Consonni. 2010. *Molecular Descriptors for Chemoinformatics*, R. Mannhold, H. Kubinyi and G Folkers (Eds.), Vol. 1, *Methods and Principles in Medicinal Chemistry*. Weinheim, Germany: Wiley-VCH.

Trevino, V., and F. Falciani. 2006. GALGO: An R package for multivariate variable selection using genetic algorithms. *Bioinformatics*, 22 (9):1154–1156.

Trevor, H., and E. Brad. 2013. lars: Least angle regression, Lasso and forward stagewise. http://CRAN.R-project.org/package=lars.

Tropsha, A. 2003. Recent trends in quantitative structure-activity relationships. In *Burger's Medicinal Chemistry and Drug Discovery*, D.J. Abraham (Ed.), pp. 49–76. Hoboken, NJ: John Wiley & Sons.

Tropsha, A. 2010. Best practices for QSAR model development, validation, and exploitation. *Molecular Informatics*, 29 (6–7):476–488. doi:10.1002/minf.201000061.

Tropsha, A., and A. Golbraikh. 2010. Predictive quantitative structure-activity relationships modeling. In *Handbook of Chemoinformatics Algorithms*, pp. 211–232. Boca Raton, FL: Chapman and Hall/CRC.

Tropsha, A., A. Golbraikh, and W.J. Cho. 2011. Development of kNN QSAR models for 3-arylisoquinoline antitumor agents. *Bulletin of the Korean Chemical Society*, 32 (7):2397–2404. doi:10.5012/bkcs.2011.32.7.2397.

Vainio, M.J., and M.S. Johnson. 2005. McQSAR: A multiconformational quantitative structure—Activity relationship engine driven by genetic algorithms. *Journal of Chemical Information and Modeling*, 45 (6):1953–1961. doi:10.1021/ci0501847.

Varnek, A. 2011. Fragment descriptors in structure-property modeling and virtual screening. In *Methods in Molecular Biology*, pp. 213–243. Totowa, NJ: Human Press.

Vasanthanathan, P., O. Taboureau, C. Oostenbrink, N.P.E. Vermeulen, L. Olsen, and F.S. Jørgensen. 2009. Classification of cytochrome P450 1A2 inhibitors and noninhibitors by machine learning techniques. *Drug Metabolism and Disposition*, 37 (3):658–664. doi:10.1124/dmd.108.023507.

Verma, R.P., and C. Hansch. 2005. An approach toward the problem of outliers in QSAR. *Bioorganic & Medicinal Chemistry*, 13 (15):4597–4621. doi:10.1016/j.bmc.2005.05.002.

Vilar, S., G. Cozza, and S. Moro. 2008. Medicinal chemistry and the molecular operating environment (MOE): Application of QSAR and molecular docking to drug discovery. *Current Topics in Medicinal Chemistry*, 8 (18):1555–1572.

Wagener, M., J. Sadowski, and J. Gasteiger. 1995. Autocorrelation of molecular surface properties for modeling corticosteroid binding globulin and cytosolic Ah receptor activity by neural networks. *Journal of the American Chemical Society*, 117 (29):7769–7775. doi:10.1021/ja00134a023.

Wang, T., M.B. Wu, J.P. Lin, and L.R. Yang. 2015. Quantitative structure-activity relationship: promising advances in drug discovery platforms. *Expert Opinion on Drug Discovery*, 10 (12):1283–1300. doi:10.1517/17460441.2015.1083006.

Wiecek, M., T. Kottke, X. Ligneau, W. Schunack, R. Seifert, H. Stark, J. Handzlik, and K. Kiec-Kononowicz. 2011. N-Alkenyl and cycloalkyl carbamates as dual acting histamine H3 and H4 receptor ligands. *Bioorganic & Medicinal Chemistry*, 19 (9):2850–2858. doi:10.1016/j.bmc.2011.03.046.

Wiener, H. 1947. Structural determination of paraffin boiling points. *Journal of American Chemical Society*, 69 (1):17–20.

Wishart, D.S., C. Knox, A.C. Guo, D. Cheng, S. Shrivastava, D. Tzur, B. Gautam, and M. Hassanali. 2008. DrugBank: A knowledgebase for drugs, drug actions and drug targets. *Nucleic Acids Research*, 36 (Database issue):D901–D906. doi:10.1093/nar/gkm958.

Wishart, D.S., C. Knox, A.C. Guo, S. Shrivastava, M. Hassanali, P. Stothard, Z. Chang, and J. Woolsey. 2006. DrugBank: A comprehensive resource for in silico drug discovery and exploration. *Nucleic Acids Research*, 34 (Database issue):D668–D672. doi:10.1093/nar/gkj067.

Xiao, N., D.S. Cao, and Q.S. Xu. 2016. enpls: An R package for ensemble partial least squares regression. *ArXiv Preprint*, 1–12.

Xue, Y., Z.R. Li, C.W. Yap, L.Z. Sun, X. Chen, and Y.Z. Chen. 2004. Effect of molecular descriptor feature selection in support vector machine classification of pharmacokinetic and toxicological properties of chemical agents. *Journal of Chemical Information and Computer Sciences*, 44 (5):1630–1638. doi:10.1021/ci049869h.

Yee, L.C., and Y.C. Wei. 2012. Current modeling methods used in QSAR/QSPR. In *Statistical Modelling of Molecular Descriptors in QSAR/QSPR*, pp. 1–31. Weinheim, Germany: Wiley-VCH Verlag GmbH & Co. KGaA.

Yoo, C., and M. Shahlaei. 2017. The applications of PCA in QSAR studies: A case study on CCR5 antagonists. *Chemical Biology & Drug Design*. doi:10.1111/cbdd.13064.

Yousefinejad, S., and B. Hemmateenejad. 2015. Chemometrics tools in QSAR/QSPR studies: A historical perspective. *Chemometrics and Intelligent Laboratory Systems* 149, Part B: 177–204. doi:10.1016/j.chemolab.2015.06.016.

Zefirov, N.S., and V.A. Palyulin. 2002. Fragmental approach in QSPR. *Journal of Chemical Information and Computer Sciences*, 42 (5):1112–1122.

Zhong, M., X. Nie, A. Yan, and Q. Yuan. 2013. Carcinogenicity prediction of noncongeneric chemicals by a support vector machine. *Chemical Research in Toxicology*, 26 (5):741–749. doi:10.1021/tx4000182.

Zhou, J.Z. 2011. Chemoinformatics and library design. *Methods in Molecular Biology*, 685:27–52. doi:10.1007/978-1-60761-931-4_2.

Index

Note: Page numbers followed by f and t refer to figures and tables respectively.

Y

Z